Inequality

Inequality

A Genetic History

Carles Lalueza-Fox

The MIT Press
Cambridge, Massachusetts
London, England

The MIT Press would like to thank the anonymous peer reviewers who provided comments on drafts of this book. The generous work of academic experts is essential for establishing the authority and quality of our publications. We acknowledge with gratitude the contributions of these otherwise uncredited readers.

This book was set in Stone Serif and Stone Sans by Jen Jackowitz. Printed and bound in the United States of America.

Library of Congress Cataloging-in-Publication Data

Names: Lalueza i Fox, Carles, 1965- author.
Title: Inequality : a genetic history / Carles Lalueza-Fox.
Description: Cambridge, Massachusetts : The MIT Press, [2022] | Includes
    bibliographical references and index.
Identifiers: LCCN 2021011599 | ISBN 9780262046787 (hardcover)
Subjects: LCSH: Social status—Health aspects. | Equality—Health aspects.
    | Human genetics—Social aspects. | Human population genetics. | Human
    evolution. | Sociobiology.
Classification: LCC RA418.5.S63 .L358 2022 | DDC 362.84—dc23
LC record available at https://lccn.loc.gov/2021011599

10 9 8 7 6 5 4 3 2 1

Contents

Preface

History is a nightmare from which I am trying to awake
—James Joyce, *Ulysses*

When I was a child, there were many history books at home that my father had bought—but that sadly, he never had the time to read. Even the street where we lived, in the Gothic Quarter of Barcelona, was made of history, with most of its buildings dating back to the Middle Ages. Ancient history was a matter of conquerors, dramatic quotes, bloody battles, and deaths. But the anonymous folks, the common people—the bulk of the population—were never even mentioned. Although my career gravitated toward evolutionary biology, history always remained my main intellectual interest. (One might argue, of course, that the two disciplines, along with astronomy, share one element: the time dimension.)

After years of working on Neanderthal genetics, I grasped that the novel DNA sequencing technologies could help us explore the recent human past using a new, multidisciplinary approach that integrated genetics, archaeology, anthropology, and even linguistics. In 2014, I led the retrieval of the first European forager genome from an eight-thousand-year-old skeleton and the next year the first early farmer genome from the Mediterranean.[1] In subsequent years, I continued this research with more projects, examining different archaeological horizons, mainly in collaboration with Harvard researcher and pioneer in the field David Reich. And as we moved closer to the present, we also started incorporating information from historical sources into the general picture—and sometimes challenging it. A common finding from these studies is that migration, and not just the propagation of ideas, was a prevalent phenomenon of the past and that modern human populations were in fact shaped by successive layers of different genomic ancestries associated with these past population movements.[2]

One day, in a casual conversation about my work, my wife said that I tended to look into the past through a man's eyes and that the history of humanity—in fact a long road of suffering and discrimination, still going on for many—included women, no less than half the world's population. And she was right; though women are largely ignored in the old history books, they crucially bear each new generation of humankind. Just think how differently a legendary tale like the kidnapping—and by modern standards, subsequent rape—of Sabine women by Romulus and his companions in the early history of Rome, abundantly represented in art in a rather heroic manner, might be told from the female rather than the usual male perspective.

I realized that directly or indirectly, the new genetic studies were in fact revealing numerous layers of inequality in past societies, from the potential

Figure 0.1
The Intervention of the Sabine Women, painted by Jacques-Louis David (1748–1825) in 1799 (on display at the Louvre Museum, Paris). This episode, recounted by Livy and Plutarch, was frequently depicted since the Renaissance as an inspiring example of ancient Roman heroism.

gender biases that we were unearthing in those migrations to the social structures implemented to maintain such inequalities as well as correlating wealth and social status with sex, kinship, and ancestry in cemeteries.[3] Powerful men of the past could have more offspring, from different women, than contemporaneous men that were left with no children—or with children who had less chance to survive.

Some recent studies set out to analyze the genetic composition of African slaves, and in conjunction with genomic analysis from recently admixed modern populations—notably those of the Americas—it was possible to reconstruct unequal reproductive patterns. Again, if we shift our point of view, certain anecdotes of the past, such as that George Washington's dentures were possibly made from teeth pulled out of the mouths of Black slaves, are all the more shocking and quite understandably have generated a broad range of reactions.[4]

Crucially, such patterns of inequality all left genetic marks that can be recognized in the genomes of ancient and modern human populations. Whenever I looked at a new genetic study, I saw new evidence of inequality and discrimination in different times and periods, and I decided to write this book to speak for those past figures who suffered the consequences. A number of baffling ideas came out of these observations. To mention but a few: those who sustained inequality in the past and had more offspring are more likely to be among our genetic ancestors, and if wealthy men could mate with multiple women and this was a prevalent pattern, then clearly women have contributed more than men to modern human genetic diversity.

Philosopher Walter Benjamin was right when he said, "It is a more arduous task to honor the memory of anonymous beings than that of famous persons."[5] But with genetic data, this task is now possible to achieve. The main observation is that history—the history of heroic acts, wars, and conquests—has indeed been a tale of inequality that modeled the genomes of humankind.

That said, inequality is not just a curiosity of the past. When I started this book, I predicted that inequality would differentially influence mortality in the current COVID-19 pandemic, and a few weeks later, my hunch was confirmed. Inequality is entangled in our genomes, but it also casts a long shadow over the future of society. We'll need to decide sooner than later how we want to face it.

1   The Age of Inequality

Inequality was the unalterable law of human life.
—George Orwell, *1984*

We live in an age of growing inequality, and to such an extent that the concept has become a social concern. In Google searches, the word "inequality" has been on the rise, especially since the year 2004. In a database of millions of books published since 1800 on Google Ngrams, the frequency of the word "inequality" remains almost undetected from 1800 to 1960; thereupon it surges continuously. In 2013, President Barack Obama said that income inequality is "the defining challenge of our time," and one year later, Pope Francis stated, "Inequality is the root of social evil." According to Oxfam's 2020 report, the 2,153 richest people on the planet have amassed more wealth than 60 percent of humankind, and according to Swiss bank UBS, the global billionaire wealth climbed to a new record of $10.2 trillion during the peak of the COVID-19 pandemic, between April and July 2020.[1]

Parallel to a general public interest in present-day disparity in incomes there has been an academic interest in understanding the causes and consequences of its increase. In this context, the success of the 2014 book by the French economist Thomas Piketty, *Capital in the Twenty-First Century*, might not be that surprising. According to the author, at the global level, 1 percent of the population has more than twice as much wealth as 6.9 billion people, while 70 percent of the population altogether accounts for just 3 percent of the world's wealth. In the United States, the richest 10 percent possess more than 72 percent of the national wealth. In countries such as France, Germany, Great Britain, and Italy, the richest 10 percent possess around 60 percent of the national wealth, while the poorest half the population have a low income, collectively owning just 10 percent. In countries as diverse as China, Germany, Egypt, and Thailand, income inequality in the year 2000 looked strikingly similar to what it was in 1820. And places like Mexico and Brazil are more unequal now than when military and political leader Simón Bolivar was alive.[2] For those hoping to make the world a better place to live, it is a startling revelation.

If we consider the world as a single population and work with the Gini coefficient—named after the statistician and sociologist Cirrada Gini, who proposed it in 1912—which ranges from 0 (all people are equal and have the same income) to 100 (a single person owns everything), we discover, according to Piketty, that the figure has increased from 43 in 1820 to 66 in the year 2000. But the increase has not been steady across this period; in fact, it has been growing sharply since 1980, especially in places like the United States, where in 2012 it was at a staggering 85.2 out of 100 on the scale.[3] Despite criticisms of the use of this index and controversies surrounding Piketty's interpretations, one thing that we can all agree on is that no one can predict the limit to this increase in the world's inequality.

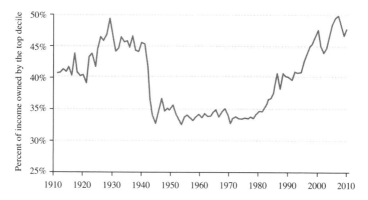

Figure 1.1
Income inequality in the United States over the last one hundred years, shown by the percent of the country's income that is owned by the richest 10 percent of the population. Despite a significant reduction, mainly associated with World War II, there has been a continuous increase over the last forty years. *Source*: T. Piketty, *Capital in the Twenty-First Century* (Cambridge, MA: Belknap Press of Harvard University Press, 2014).

Perhaps in no other nation has the growth in inequality been so evident or studied in such detail as in the United States. Throughout the twentieth century, and despite continuous economic growth, the wealthiest have accumulated much more than ever before, while the bottom 50 percent of national income shares have remained stagnated or even gotten poorer in some places.[4] Between 1979 and 2007, the real income for the wealthiest 1 percent of the nation tripled, while that of the middle class increased by only 25 percent (and even so, due to an increase in working hours and labor mass). In Piketty's words, income inequality derived from work in the United States "is probably higher than in any other society at any time in the past, and anywhere in the world including societies in which skill disparities were extremely large."[5]

Such social factors as the aging of the population, the globalization phenomenon, and migratory movements from poor to rich countries as well as the growing threat of climate change and the current COVID-19 pandemic indicate that inequality will likely continue to grow, mainly because state resources must be shifted from diverse redistributive policies to basic public services. Also, new technologies and industrial automation suggest that the income gap between the qualified and unqualified workforce will continue to increase in the near future. In his 2018 book *Enlightenment Now: The Case for Reason, Science, Humanism, and Progress*, the Canadian evolutionary psychologist Steven Pinker emphasizes the fact that at the planetary level, inequality *between* countries is lower now than at any time in the past and that there has been a rise in the number of poor people becoming middle class in developing countries—cause for optimism in his view.[6] According to Oxford University economist Ian Goldin, however, globalization has also led to a dramatic increase in systemic risks, including global warming and the degradation of the world's ecosystems.[7] Moreover, Pinker downplays the fact that inequality is effectively growing within each country (this is one of the reasons that he, along with others such as Hans Rosling, are sometimes dubbed the "New Optimists").[8] And the maxim that differences in technological societies will be primarily meritocratic is likely false; the economy will be dominated by inherited wealth.[9] The ongoing robotization of production, for instance, means that inequality will continue to rise, at least in the near future, and that large segments of society—those with fewer qualifications—will remain jobless.

The essential point of Piketty's work is that nowadays, the return on financial capital is much higher than what is yielded by economic growth (equivalent to human capital)—a process that in turn enables wealth to accumulate faster. The market economy, unrestrained by any form of regulation, contains divergent forces that favor a much quicker increase in wealth than salaries and production can provide, and in the absence of a solution—here Piketty proposes a global tax on capital—this trend will continue to increase inexorably. Although a number of economists have jumped into the arena criticizing aspects of the French researcher's thesis from his methodology to his conclusions and proposals, there seems to exist a general consensus within academic circles that inequality—understood as differential access to wealth—is effectively larger now than before the 2008 crisis. Also, some researchers have argued that this long trend of rising inequality can be considered an unavoidable epiphenomenon of the progress that civilization has experienced since the Middle Ages; that is, inequality is the toll to pay for progress.

Whatever the magnitude, the existence of such rampant inequality is no trivial matter, and beyond moral considerations, has an important impact on health, prosperity, and even life expectancy throughout society. Several studies have shown that citizens from the most disadvantaged strata in a highly unequal society have a higher chance of becoming obese, sick, or psychologically afflicted, and are more likely to populate its prisons. We have now learned that they are more likely to die during a pandemic too. These results, beyond the negative consequences at the personal level, mean that inequality undermines the productivity of any economy due to the large costs of maintaining public health systems as well as security enforcement. And in the long term, the situation will probably favor the emergence of authoritarianism to ensure the persistence of current inequality—or rather paradoxically, to claim to fight against it—as well as the parallel advent of political extremisms that could eventually destabilize the system. Some observers have warned of signs associated with both phenomena—the former in the United States and within some Asian powers, and the latter in some European countries.

If inequality has grown to become a distinctive sign of our society, what can be done to fight it? To Stanford University professor Walter Scheidel, there are four mechanisms for correcting inequality that have worked historically and are much more effective than those suggested by economists:

war, revolution, state collapse, and deadly natural disasters—including pandemics.[10]

In the first case (war), conflicts must be paid for through an increase in taxes on the wealthiest (in some cases, as in both world wars, such taxes reached 90 percent) as well as social mechanisms for the underprivileged classes to compensate for the cost in human lives. In the second case (revolution), social movements such as the Russian and Chinese Revolutions erased the ruling social class in a quick and effective way. Something similar occurred in the United States during the Great Depression. But revolutions aren't always effective in modifying inequality; according to Piketty, the French Revolution only represented a small drop in the country's inequality levels, with the wealthiest 1 percent in Paris owning 55 percent of all private property in 1780 versus 49 percent in 1810 under Napoléon. Meanwhile, the third case (state collapse) is associated with a number of phenomena. For example, the fall of the Western Roman Empire was followed by a large-scale decrease in inequality, yet this was achieved through the undesirable process of the impoverishment of the entire population. Finally, in the fourth case, dramatic natural disasters or large-scale pandemics, if they indiscriminately affect the whole society, can also decrease levels of inequality. The best known of these past pandemic events, the Black Death, which killed off between 30 and 50 percent of Europeans during the fourteenth century, rather than destroying the social order, freed the few surviving workers to look for better employers; this process triggered the end of the Middle Ages as well as the emergence of a middle class and urban population centers.[11] In fact, studies of tax records in Italy by researchers Guido Alfani and Matteo Di Tullio show that the Black Death was the only time in the past few hundred years when inequality decreased in that country.[12]

There are in addition local phenomena. In the United States, for instance, inequality experienced a decrease after the Civil War, mainly due to the emancipation of slaves in the southern states. Moreover, according to Scheidel, none of these mechanisms that have operated in the past will have the same effect on modern inequality, which is much more complex and interconnected on a global scale.[13] In conclusion, there seems to be no easy solution to this problem, and the past is not likely to help us find suitable mechanisms—unless, of course, we accept a world catastrophe as a leading force for change.

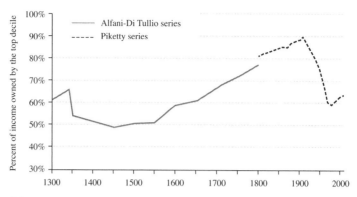

Figure 1.2
Evolution of the share of wealth of the richest 10 percent in European societies from 1300 to 2010. During this entire period the only two episodes of significant decline in inequality are related to catastrophic events: the Black Death in 1347 and subsequent years and the two world wars. The share of the richest 10 percent today is about the same as that in Europe (or Italy, at least) immediately before the Black Death. The Alfani-Di Tullio series (2019) is an average of the Sabaudian state, the Florentine state, and the kingdom of Naples (before 1600, only information from the first two is available). The Piketty series (2014) is an average from France, the United Kingdom, and Sweden. *Source*: G. Alfani and M. Di Tullio, *The Lion's Share: Inequality and the Rise of the Fiscal State in Preindustrial Europe* (Cambridge: Cambridge University Press, 2019) and T. Piketty, *Capital in the Twenty-First Century* (Cambridge, MA: Belknap Press of Harvard University Press, 2014).

If we would like to fight inequality, however, a pertinent question to ask from the field of social psychology is to which extent people view it as a problem that needs to be addressed; for example, a recent poll about income inequality in the United States showed that less than half of North Americans consider it a serious concern.[14] Researchers on social psychology and moral economics have explored the innate bases of social differences, finding that in many cases people seem to prefer societies with inequalities (or a certain amount of inequality) over equal ones.[15] Some researchers think that this conclusion can be explained because what really seems to worry people is not inequality per se but instead economic unfairness—different concepts that sometimes get mixed up. This trend seems to be universal (though some studies indicate that people don't realize the magnitude of inequality in our current society). Various transcultural tests conducted on children suggest that humans do favor fair over merely equal distribution, and that in situations of conflict between justice and equality, people tend to prefer fair inequality over unfair equality.

In social games performed under controlled conditions, the perception of justice is prevalent in all human interchanges, even beyond rational considerations. Unfair players are often punished, even where their opponent forfeits all potential benefits. An innate altruism, influenced by considerations of social prestige, seems to be a common conclusion in such experiments.

The most common of these social experiments, the "ultimatum game" (designed in 1982 by researchers from the University of Cologne in Germany) concludes that a perception of fairness is a fundamental element in human exchanges, even more than rationality.[16] In the game, two players share a certain amount of money (say, ten dollars); one makes a single, final offer to the other, who may accept or reject it. If rejected, both players are left with nothing. The experiment has been repeated many times with different players and communities, and the obtained results are always similar: obviously unfair proposals (such as a distribution of nine to one) are rejected, even those that yield a net gain for the second player. (We might think, from a purely rational point of view, that one dollar in profit is better than nothing at all.) Most transactions accepted are close to what is perceived as fair—five to five—while offers below two are generally rejected by the second player. If the proposal is not fair, it is preferable to punish the greedy player, leaving them with nothing. Brain scanners performed on the second player during the decision process showed an activation of the dorsal striatum, where reward perceptions reside in the brain. The satisfaction experienced in teaching greedy opponents a lesson seems to dominate the player's visceral response. Of course, there are quite a lot of intercultural variations in these fairness judgments, and yet what seems to underlie them is the existence of a general psychological mechanism adapting to different conditions in which the second player never behaves like a rational economic maximizer.[17]

That's not all; unbalanced offers are usually not proposed because the first player knows that a markedly unfair offer will be rejected and prefers not to run the risk of being punished. Obviously if this is the case and the first player is making a strategic decision, then we should believe that offers are rationally constructed, which is what conventional economic theory says. That said, other researchers have suggested an alternative, altruistic explanation to this bias: those who make an offer prefer to set a more equitable partition from the outset.

In a variant of the game, the first player is a machine and the second one is aware of this circumstance. In this case, when the offer is unfair, the other player—the human being—accepts it because there would be no point in teaching a moral lesson to a machine.

In the "dictator game," there is a variant: the player who receives the offer is a passive subject and cannot reject it.[18] Despite the potential impunity of the first player, it could be seen that the second player ends up with amounts smaller than those from the ultimatum game, but nevertheless higher than zero. It is difficult to know whether or not the altruistic component is in this case overtaken by matters of social prestige. In some variants of the game, it is explained to player one that their opponent will never know who they are—and therefore no one will ever know that they are greedy—and then it turns out that this person makes more unfair transactions, protected by this feeling of social impunity. Some people have argued that in real life, and with much larger amounts of money at risk, the other players would behave more rationally, insisting that such economic games are not representative of human realities, which are much more complex and unpredictable, and subject to more intangible personal variables and special circumstances.

While such criticisms may be fair, a common conclusion that can be drawn from all these experiments is that economic transactions, rather than cold, rational activities, are largely based on emotional reactions in human societies (something that marketing experts are well aware of). This is important because our civilization has been built on the basis of commercial exchanges, especially since the development of farming—and the possibility of having food surpluses to trade. Integration in a social context is thus a determinant in the generosity of economic transactions.

Surprisingly, the emotional component is not exclusive to humans. Experiments undertaken with capuchin monkeys showed that they could tell when something was fair or not by having some of them watch others being rewarded with a grape instead of a less tasty cucumber. In the experiment, monkeys were taught to exchange a token for a grape; if they were in isolation, they felt satisfied with either a grape or cucumber, but in the presence of more monkeys, they become quite irate if they found themselves at the short end of an inequitable distribution of rewards.[19] I suppose many non-monkey readers will identify with this feeling.

Based on all these observations, some researchers suggest that an idea of equity—defined as a benefit according to relative contribution—exists because it has been naturally selected in humans as the best psychological strategy to be integrated into a cooperative environment in which humans evolved.[20] (By the way, inequity and inequality are different concepts; the former is an unfair or unjust state of being, and the latter is an imbalance that may or may not depend on inequity.)

And yet we all would agree that common experience indicates that inequity is ubiquitous in our societies. If both fairness and equity—as well as inequality and inequity—are so prevalent, they might have a biological basis; consequently we might ask ourselves if they can be considered to some extent unavoidable.

And following this, one might hypothesize as well that inequity is somehow related to innate aggressiveness and dominant behavior, which in the natural world could result in differences in strength and body size.[21] (Interestingly, humans show a level of sexual dimorphism—understood as systematic physical differences between both sexes—that could also be associated with ancestral differences in aggression levels between males and females.)

Can we study aggression from an evolutionary perspective? When one of our closest relatives among the great apes, the common chimpanzees, are studied, researchers describe solidarity among fellow primates, but also aggression and competition associated with the highly hierarchical structures of chimpanzee groups; wars are even waged between rival groups, and dominant males tyrannize female members. (In contrast, the bonobos or pygmy chimpanzees use sex to maintain a peaceful social structure.) According to British primatologist Richard W. Wrangham, two different types of aggression may be distinguished here: proactive (similar to offensive, planned, controlled, or cold aggression) and reactive (defensive, emotional, or hot aggression). Cold, tactical proactive aggression is more pronounced in humans as compared to other primates, while hot, reactive aggression is more frequent among chimpanzees. The low propensity for reactive aggression in humans could be constrained, according to Wrangham, by a tendency toward extensive within-group tolerance and even cooperation.[22]

Cold, proactive aggression might have been favored among early hominins if altruism genes in the group increased the chances of winning conflicts against neighboring groups by enhancing intragroup cooperation bonds. For this mechanism to operate effectively, genetic differentiation between hunter-gatherer groups had to be sufficiently high in the past—something that seems to be at odds with present genetic evidences for widespread admixture among African forager groups prior to the advent of farming.[23] Proactive aggression, however, also existed within groups, with the possibility of enforcing norms through such measures as capital punishment that might have selective consequences. Some researchers disagree with Wrangham's two categories of aggression, arguing that more than two basic types can be distinguished.[24] Notwithstanding this controversy, the notion that aggression might have been a driving force in our evolution remains powerful.

One potential source of evidence for the study of the evolutionary trajectory of aggression is the fossil record, where we can study what are in all likelihood our closest extinct relatives, the Neanderthals. We can speculate that if there were behaviors that helped our species to survive, then differences in the outcome of both modern humans and Neanderthals might have resulted from such behaviors. We find evidence of compassion among Neanderthals, with disabled individuals surviving for years under the group's care. But we see signs of violence too, with an abundance of trauma and cannibalism. These somewhat contradictory findings are in fact quite humanlike, and we can recognize ourselves in them, both for good and evil. Nevertheless, the evidence of aggression and cooperation doesn't tell us how prevalent the two types of behaviors were, much less what kind of aggression was occurring.

From a demographic point of view, we now know that Neanderthals experienced a long process of population decline—well before their encounter with modern humans—while our species underwent a continuous expansion after the out-of-Africa migration—or migrations—that took place around sixty-five thousand years ago. It can be demonstrated from genetic data that there were various episodes of interbreeding not only between Neanderthals and Homo sapiens but also between both and yet another archaic hominin group from Asia, the Denisovans.[25] Still, no sign of gene flow from our ancestors into the latter Neanderthals (those dated to the potential contact period, between forty to fifty thousand years ago) has been detected so far. It seems that modern humans were able to tolerate hybrids but not vice versa. It could be that hybrids were rejected from Neanderthal groups—or maybe even killed—yet their absence could just be the consequence of a greater demographic fragmentation or even existing genetic differences. In fact, the Y chromosome sequence of some Neanderthal individuals showed specific mutations in three genes that could account for an incompatibility of hybrids with a Neanderthal father only.[26] Whether these patterns represent fundamental differences in aspects such as social structure or aggression, or are just a reflection of biological incompatibilities, is something that remains to be explored. But so far there is no evidence of Neanderthal–modern human hybrids living in the small, endogamous, and kin-structured Neanderthal groups.[27] One might be tempted to think that accepting admixed people as equals could hold clues to our success as a species, at least in these distant periods.

Yet if proactive aggression was a driving force in the successful expansion of Homo sapiens across all continents, is this a trait, we might ask ourselves, that has been increasing in tandem with our current demographic explosion? Did we become more abundant by becoming coldly aggressive? In his 2011 book *The Better Angels of Our Nature: Why Violence Has Declined*, Pinker argues that despite what people may perceive and the dramatic statistics from the two world wars, the number of deaths in violent conflicts (corrected for the rise in population) is continuously decreasing.[28] Moreover, some researchers reported that forager groups suffer from higher homicide and war occurrence rates than previously assumed.[29] This conclusion has been criticized by some historians and ethnographers alike, partly because of the inherent difficulties in estimating the number of deaths from past conflicts and partly because recent ethnographic studies suggest that the basic pattern of interaction among forager groups is to get along with neighbors rather than attacking them.[30] Meanwhile, economist and essayist Nassim Taleb has also questioned Pinker's idea that humanity is entering a period of "lasting peace," as he calls it. Taleb argues that world-scale armed conflicts do not follow a predictable pattern because of the long interarrival time existing between conflicts and therefore Pinker's results can be considered a "statistical illusion."[31]

The conclusion of this controversy is that we do not know if past forager groups were peaceful or inclined toward violence. But beyond the academic debate over the past and future of war, one thing seems evident: rather than arising out of the sedentary lifestyle of the first farming communities, violence and conflicts are part of the emergence of human nature, as are their counterparts, altruism and solidarity. Despite some opposite claims, the former is not just a recent phenomenon; there are some examples of deliberate interpersonal violence and resulting lethal trauma in fossils from the Middle Pleistocene, such as in a skull from the famous Sima de los Huesos site in Atapuerca (Spain) that received multiple, deadly blows.[32] The issue of war mortality before the Pleistocene is still hotly debated, though.

Summing up, there are different approaches to the study of innate human behavior, and most seem to conclude that cooperation as well as competition—including aggression—are prevalent within and between social groups.[33] It can be hypothesized, perhaps controversially, that inequity is rooted in this innate behavior and thus might never be eradicable. By the same token, inequalities deriving from unfair conditions, a kind of cold proactive structural violence, might also be impossible to eliminate. Therefore a crucial question for evolutionary biologists is whether inequality does in fact have a biological basis, rooted in proactive violence. It is likely that if we are to understand inequality, we must understand violence and its evolution.

Beyond this debate and irrespective of its nature, we're still interested in analyzing not the basis but rather the consequences of this inequality over the genetic structure of human populations, both past and present. How we can measure it?

The study of the accumulation of resources—that is, wealth—is key to understanding inequality in premodern societies. Such study requires one to quantify the different classes of wealth—material, embodied, and relational—and the mechanisms of their transmission across generations (though some argue that these different classes are not independent of each other). Material wealth refers to those goods—land, livestock, household, or objects—that can be possessed and accumulated; embodied wealth refers to health, immunity, physical strength or body weight, knowledge, and practical skills stored in human bodies; and relational wealth consists of the individual social links that derive from social status or family ties.[34] Transmission of material goods between generations, usually referred to as "inheritance," is a defining human trait rarely found in nature.[35] Some studies on modern African agricultural societies suggest that material and relational wealth are likely more important in the emergence and persistence of inequality than embodied wealth, and that ownership rights in land and stock are more significant factors in determining the levels of inequality than the use of domesticated animals itself.

The study of inequality in all its aspects can to some extent be based on the analysis of quantitative data from the foraging, agriculturalist, and pastoralist groups of recent times. For most of our evolutionary history, humans have been foragers (or hunters and gatherers). Due to fluctuating and uncertain resources, foraging groups are constantly on the move, and hence cannot accumulate goods or even food. Therefore these foragers are typically described by ethnographers and cultural anthropologists as having formed more egalitarian societies—including sex equality—than agriculturalist and pastoralist groups.[36]

But how can we understand—if we can—the extent and level of potential inequality in the distant past? Sustained economic inequalities usually leave archaeological signatures, although they can be difficult to interpret with our eyes, which are those of a rich society. Their generalized absence (in the form of rich funerary assemblies, for instance) prior to the Neolithic period has been interpreted as evidence of scant economic differentiation between and within foraging groups, with the possible exception of those occupying abundant fishing sites.[37] As we have seen, observations based on modern foraging groups have lent support to this interpretation, but it is difficult to know how well present-day anthropological data can be extrapolated to the past. Are modern forager groups comparable to those of the past?

Actually, some rather elaborate graves have indeed been found dating as far back as the foraging groups of the Upper Paleolithic—long before either permanent settlements or agriculture—that bear marks of symbolic behavior such as ochre soil scattered over an interred body, or stone tools or animal bones that seem to have been deposited alongside. The most ornate example of such burials is in Sunghir near Vladimir (Russia), where an adult male and two juveniles dated to about thirty-four thousand year ago were excavated between 1957 and 1977. There were about three thousand mammoth ivory beads on the adult skeleton (originally sewn onto their clothing), plus twelve pierced fox canines and twenty-five mammoth ivory armbands; all these items of personal decoration required hundreds of hours of manufacturing. But the juveniles, approximately ten and twelve years old, and buried head to head, were even more spectacularly decorated, with about five thousand mammoth ivory beads each.[38] What do these mean in terms of hierarchy, group membership, and social structure? Are they signs of a certain status among these foraging groups, or simply well-preserved burials in particular contexts?[39] Also, numerous examples of the Paleolithic Venus figurines—representing women with exaggerated anatomical traits related to fertility—have been found across Europe, but do they mean that women were held in higher regard, perhaps due to their involvement in fertility rituals, than in subsequent periods, or do they just reflect hopes for survival in harsh times?[40] Again the problem is, What can be deduced and generalized from archaeological findings in terms of social stratification and gender roles?[41]

The difficulties in understanding potential signs of inequality in the fossil and archaeological records can be easily grasped by looking at the most basic data that one can hope to get from the past: the gender of the people involved. Let's look, for instance, at the famous Upper Paleolithic triple burial dating from around thirty-one thousand years ago from Dolní Věstonice in the Czech Republic. The three individuals in this burial (labeled DV13, DV15, and DV14 by archaeologists) are intriguing for a number of reasons. DV13—which corresponds to a male teenager—had a thick wooden pole driven through his hip and was found with his hands on the pubic region of the central individual, DV15. Flanking DV15, a second male teenager, DV14, was buried facedown.[42] The central individual, who obviously had a preeminent status, had usually been characterized as female, although the real sex was controversial due to a pathological condition affecting the curvature of the spine, pelvis, and teeth.[43] Until recently, the biological sex of skeletal remains had to be assessed via morphological indications as, on average, males have larger and more robust bones than females, and in this case it was unclear. One popular interpretation was that the middle individual was a high-status female who died during childbirth—because they had ochre over their pelvic bones—and the man on the right was possibly a kind of shaman who was executed for failing to save her life. Our view of this triple burial might alternatively be considered the result of a tragic love triangle with the central character being female or all three bodies being male (which would not of course rule out the possibility of a love triangle). Genomic data generated from the three individuals and published in 2016 eventually revealed that all three were in fact male, thus ruling out some of the previous interpretations and changing the way that we perceived this remarkable burial—and also the potential social relationships within the group.[44]

Not only the genetic sex can now be unveiled, but the kinship. The genetic analysis of the Sunghir skeletons revealed, for instance, that the two immature individuals buried together were not—despite previous suggestions—family related.[45] These are but two examples of how genetic data can shed new light on the study of the past, resolving even for seemingly unresolvable mysteries.

In the last five thousand years, signs of inequality in funerary contexts have increased, yet the paucity and gaps in the archaeological record impose restrictions on the conclusions that can be drawn from different times and periods. Apart from the discovery of spectacular burials of wealthy people, it is difficult to really grasp what the archaeological record has to tell us about how unequal those societies actually were.

Even in recent history and Western societies, we have some evidence to suggest a surprising connection between genetics and social inequality. Some recent studies have exposed the persistence of social status—or to put it another way, a low level of social mobility—existing even in a society like Sweden, considered a model of social democracy. An economist from the University of California at Davis, Gregory Clark, surveyed for surnames and income among present-day Swedes. The underlying idea was that no new aristocratic houses were created after the seventeenth century, and all existing by then have been recorded in the House of Nobility (or Riddarhuset); noble surnames among the general population had to have been derived patrilineally—that is, transmitted only through male relatives—from past aristocrats. (To name just a few, Björnstjerna, Cedercreutz, Falkenberg af Sandemar, Lövenskjöl, Meijerfeldt, Oxenstierna af Croneborg, Silfverschiöl, Wallenstedt, and Wrangel.) By contrast, common Swedes tend to have the patronymic particle "son" added to the end of their surnames. Clark found that despite tens of generations, false paternity events, and of course the ups and downs of fortune, the surname was still a good predictor of social status in twenty-first-century Sweden—along with such variables as wealth, occupation, education, and even longevity.[46] This correlation between surname and status has been replicated in other Western countries, including Denmark, Ireland, and Great Britain, and even in places like China and Korea, and the conclusion is that social mobility is lower than traditionally measured and at the same levels as in preindustrial times. In general, it takes at least four hundred years for elite surnames to converge with average-income families in society. Even in today's Britain, Norman surnames (Baskerville, Darcy, Mandeville, Montgomery, Percy, Neville, Punchard, and Talbot) are overrepresented among the country's elites.[47] Why this long persistence happens is unclear; obviously there seems to be a strong transmission within families of the attributes needed for social success. Whether these are genetic or environmental, or a combination of the two, is more difficult to unravel. While Clark controversially argues for the inheritance of some of the traits that lead to social success, such as intelligence, attractiveness, self-determination, and so on, others interpret these observations as a result of well-established social networks. Clark's work has also received some methodological criticism, as surname-level income averages capture a diverse set of individual- and group-level factors, and these cannot be disentangled in a comprehensive way without additional information.[48]

Despite potential statistical flaws, it seems evident that intergenerational social mobility is lower than we might think in our societies, and this is notably due to the existence of marked assortative mating—that is, people with similar phenotypes tend to mate more often than would be expected if mating were random. Even though we share environments in daily life, small details such as spoken accents or dress can indicate one's social status, wealth, or educational level to a high degree of precision. Whether conscious or unconscious, assortative mating likely helps perpetuate past social differences.

And even if we were to conclude that such structures are not as rigid and endogamous as those seen in societies like India, it is also obvious that people tend to marry others of the same socioeconomic status. This tendency has increased, paradoxically, with the access of women to higher education and the job market. (It might be partially correlated as well with the fact that married couples were much younger one generation ago, before they had attained their final educational and job level.) To put it simply, some decades ago doctors married nurses, and now doctors marry other doctors (in United States, approximately one in four women physicians are married to doctors, according to a report from the American Medical Association). In a study carried out in 2011 within member states of the Organization for Economic Cooperation and Development, it was observed that about 40 percent of couples had similar salaries, against a 33 percent figure just twenty years before.[49] If the previous tendency of couples with dissimilar incomes had held, present-day inequality would instead be around 25 to 30 percent lower. And it is likely that this is happening everywhere; in Spain, for example, over 40 percent of the children whose mothers have a PhD are fathered by men who have a doctorate too.[50] This cultural trend contributes to an increase in inequality, along with the aforementioned weight of one's past family status.

One way or another then, unequal societies can clearly be linked to biological factors: wealthy people can raise and feed more children as well as live longer than ordinary people. (Interestingly, this trend seems to be reversed today, with wealth and education inversely correlated to fertility, although it is unlikely that this was the case in the past.) Where such wealth can be effectively inherited, children will also have more children over several generations because they are more resilient in the face of economic turmoil. This will have consequences for the future of that population because each generation hinges on the fertility dynamics of the previous one. We need not refer back to the extreme reproductive asymmetry of Ottoman sultans who had literally hundreds of children; Clark demonstrated in 2005 that in England's pre–Industrial Revolution period, the poorest had, at the moment of their death, 2.2 surviving offspring on average, while the wealthiest had 4.1.[51] If anything, the Industrial Revolution tended to increase the differences in fertility—or at least the chances of survival—between social classes. The possibility of leaving more offspring of course has genetic consequences for upcoming generations. As we belong to those subsequent generations, we can take a look back and try to understand our own genesis.

The number of our ancestors doubles with each generation (we have two parents, four grandparents, eight great-grandparents, etc.). In a few more generations, the numbers of our genealogical ancestors will rise to such vast numbers—more than the number of people who've ever inhabited this planet—that only the enormous number of interconnecting links between them all could explain this paradox. In short, many of our ancestors are shared, in the same way, for instance, that first cousins share two grandparents (two unrelated people have eight unconnected grandparents instead of the four unconnected ones that first cousins do). That is, we do not have as many different ancestors as theoretically estimated because many of them are repeated; this is quite evident in small, endogamous populations such as those found in remote islands.

Nevertheless, it is apparent that some among our genetic ancestors must be overrepresented—not necessarily among our genealogical ancestors. Those who had a disproportionately high number of descendants in one particular generation—and maybe also their descendants through subsequent generations—are expected to have contributed more than average to the genetic ancestry of modern populations. These people would then necessarily be overrepresented among our genetic ancestors, partly reversing the previous trend of genetic dilution described by population geneticists as we go back one generation after another.

While this overcontribution of some genetic ancestors is expected to happen in the natural world as well, especially in non-monogamous species—think, for instance, of polygynous species in which a single male mates with multiple females—a human-specific trait that I want to elucidate in this book is whether differences in wealth give some individuals advantages over others in reproduction, thus entangling social and genetic factors in a complex way, and effectively modifying the genomic composition of subsequent generations. Consequently, if this is in fact the case, we can safely assume that we carry a higher genomic fraction from those who benefited from inequality in the past, if only by probability. I think this in itself is an interesting thought; let's keep it rolling in our minds for a while.

To understand the past of human inequality in new ways, beyond recent historical information such as that used by Piketty, we now have new technical tools that enable the retrieval of hundreds of ancient genomes generated from hundreds of skeletons distributed in space and time. The second-generation sequencing platforms, developed after 2008, now enable the retrieval of literally billions of DNA fragments from minute dental or bone samples from up to tens of thousands of years ago. These DNA fragments are computationally mapped onto the human reference genome and can then be analyzed with genetic statistical algorithms, along with other contemporaneous ancient genomes as well as modern ones from the same geographic areas. These extremely powerful analytic tools can discern different genetic compositions derived from each individual's ancestors as well as provide solid information about kinship and sex. Without doubt, this is a genetic revolution with a continuous impact on all levels of the reconstruction of the past, and it is unlikely to conform to previous notions.

Of course, there are limitations in genetics too; DNA degrades under certain environmental conditions and over time, being less likely to survive in hot climates such as those found in most of Africa and Southeast Asia, and the samples that can be analyzed are most likely a biased representation of all those who ever lived (in fact, we will never have data on everyone who ever lived and left descendants). And sometimes we do not have the remains of past individuals for analysis either because they haven't been found or haven't been preserved in the ground.

Current genetic tools, however, are so powerful that in many cases we don't even need to have the dead. For instance, we can study a substantial genomic fraction of someone from the past without having their remains, as a recent study on the descendants of a certain Hans Jonathan demonstrates.[52] Jonathan was born into slavery on the Caribbean island of Saint Croix in 1784, and was the son of an African slave and someone of European origin, most likely the secretary of the plantation's owner, who was Heinrich Ludvig Ernst von Schimmelmann. Later on, the Schimmelmanns moved to Copenhagen and took Jonathan with them; once there, Jonathan joined the Danish Navy and fought bravely in the Napoleonic Wars. Afterward, however, Schimmelmann died and his widow had Jonathan arrested, claiming that he was her property and she wanted to sell him in the West Indies. The Danish court—despite the fact that slavery was abolished by then in the country—sentenced Jonathan to be returned to the Caribbean. Before that could happen, he escaped to Iceland in 1802, settling in a small fishing village, Djúpivogur, where he had a couple of surviving children with a local woman, Katrín Antoníusdóttir, before his death in 1827.[53] It is somehow remarkable that he was accepted in Iceland, by then one of the world's most isolated communities, which nevertheless had less racial prejudice than Jonathan faced in continental Europe. Today, there are almost 900 living descendants of Jonathan, and from the analysis of 182 of them, it has been possible to reconstruct about 38 percent of his "African" genome (that is, 19 percent of his complete genome), which shows affinities with the present-day populations of Benin, Nigeria, and Cameroon. His descendants carry a small fraction of non-European DNA, while most of their genome is overwhelmingly similar to the rest of modern Icelanders. In fact, Iceland is probably one of the most genetically homogeneous populations in the world today, and Jonathan's African genes are but a single drop in the Icelandic gene pool. That said, his African genome, even if fragmented, is still discernible because it was so different from the rest. This is undoubtedly the true genetic meaning of leaving a legacy, and this book will provide many examples of such links between past and present.

It is worth emphasizing that geneticists could have conducted the Icelandic study without any historical record about the existence of Jonathan. Nevertheless, genetics does not stand alone in the task of reconstructing human history; the fields of archaeology and anthropology also play a role in this new view of the past. For instance, wealthy people have a higher than average social status in any time and period—a fact that can sometimes be recognized at archaeological sites. Burial ritual reflects the transition of this status to the other world, and thus there is a strong relationship between graves and the existing social organization. Genetic analysis of the dead therefore holds clues to understanding the societies where they lived, and both archaeology and anthropology add significant information. We are not advocating for an independent view of past controversies based solely on the genetic evidence; we are in fact pursuing a real interdisciplinary interpretation that was previously impossible to achieve. But archaeology won't be the only tool employed; underestimating genetics, archaeological academia runs the risk of being ignored by mainstream science, which may view it as entrenched in untested, ideological hypotheses.

The various links that exist between social structure and biology remain crucial to us because they may address a number of hypotheses, usually posed by archaeology, but previously untestable, about how prevalent inequality was in the past. Thanks to the potential of ancient DNA, we can now explore how dramatically ancestral shifts associated with past migrations have unequally shaped the genomes of humankind. We can also unravel how sex biases and new social structures associated with these past population movements have implemented the genomic change, effectively influencing future generations. We can elucidate the magnitude of some of the most obvious examples of inequality in human history: those related to underprivileged classes, minority groups, and enslaved people, but those suffered by women too (by and large, the most oppressed human beings in all historical periods; it is worth remembering as well that the greatest genetic difference within our species is between males and females). And this can be done on the collective and individual level; sometimes powerful men contribute disproportionally to subsequent generations by monopolizing women, and this shapes paternal chromosomes.

Until now, skeletons in themselves could not tell us anything about the link between social status and the ancestry of their possessors, but this is changing definitively. Through their DNA, ancient skeletons have much to tell us, and we're listening to their anonymous stories of inequality and suffering, and at the same time developing a new, objective, and multidisciplinary reconstruction of human history. Amazingly, this technical revolution was unforeseen even by people like me who have been working for two decades in the field of ancient DNA. What was impossible ten years ago is now becoming mainstream. This is what is happening.

We can thus ask for the first time (we, the outcome of these entangled social and biological processes), To what extent has inequality shaped the genomes of humans?

2 Shifts in Ancestry from Past Migrations

We are all migrants through time.
—Mohsin Hamid, *Exit West*

Thanks to paleogenetics—the retrieval and analysis of ancient genomes—past migrations can now be studied by directly analyzing the people involved in them rather than relying on indirect data open to multiple and often contradictory views. In a seminal work titled *Who We Are and How We Got Here*, published in 2018, Harvard geneticist David Reich presented a worldwide reconstruction of these past migrations, the nature of which have been recently revealed based on the study of ancient DNA from hundreds of bones and teeth. He also argued that "pure" populations (i.e., those having just one type of ancestry), as imagined by Victorian era racial scholars, have no basis in reality and that present-day populations are in fact complex mixtures, in different proportions, of highly divergent populations—which in turn were likely mixtures of older, also divergent populations—that no longer exist in unmixed form.[1] The evidence for migrations being prevalent in the history of humankind is central to this book because encounters between different populations settling the same land, arguably with disparate social organizations and mutually unintelligible languages, gives opportunities for the emergence of different sources of inequality.

Genetic research on various archaeological fronts is now revealing shifts in ancestry stemming from the migrations that shaped modern human populations. Eurasia is by far the most extensively studied continent, with about 80 percent of all ancient genomes having been obtained from western Eurasian samples.[2] It makes sense, then, that this continent best represents what can be achieved in the reconstruction of past genetics and inequality. With time, we should have enough genetic data to explore past migrations and their social implications on every continent, but so far Europe remains the only place where the picture is emerging in such detail.

Thus far, two continental-scale migrations have been genetically examined in Europe: the transition from foraging to farming starting ten thousand years ago, and the arrival of the steppe nomads from the Pontic steppe five thousand years later. The impact of the former process on the genetic composition of subsequent populations is mirrored on every continent, while the latter also had dramatic consequences in Central and South Asia.

The hunting and gathering way of life that our ancestors practiced over millions of years is only residually observed nowadays. Only a few hunter-gatherer groups remain in remote areas of Asia, Oceania, Africa, and South America, and even these are generally no longer isolated groups, but are being genetically absorbed into neighboring pastoralist or agriculturalist groups of larger demographic magnitude. The vast majority of human beings have adopted a food production economy and are to a large extent descendants of pioneering farmers. We might then ask ourselves, How different were foragers from farmers, and how did this substitution process take place?

It is not a novel question. As far back as the Enlightenment, people were aware of how crucial this transition must have been for human civilization. The influential Swiss writer and philosopher Jean-Jacques Rousseau (1712–1778) propagated a naive and idealistic notion of human nature, and paradoxically, a paternalistic vision of modern hunter-gatherer human groups branded by Western colonialism. That some contemporary environmentalist movements might sustain the view that nature is good and kind is a legacy of Rousseau.

Rousseau set forth his ideas about the primeval nature of human beings in a 1754 essay titled "Discourse on the Origin and Basis of Inequality among Men." The philosopher asked himself, "How shall we know the source of inequality between men, if we do not begin by knowing mankind?"[3] For Rousseau, there are two types of inequality, one natural or physical, based on differences in age, health, strength, or intellectual qualities, and another that might be termed moral or political that is established by societal convention with people's consent. This second form of inequality consists of people having different privileges such as being wealthier or more powerful than others. Rousseau wondered if there was a correlation between the two—still a pertinent scientific question to this day.

He subsequently described the main features of human beings in the early stages of civilization and a human subjected to the harshness of the weather who wandered, naked and starving, through vast forests while defending themselves against attacks by wild animals. Despite living in the wilds of nature, compared to civilized humans, this human being was strong, resolute, and healthy. At the metaphysical and moral level, the savage—like other animals—values their freedom above all because they live in the constant immediacy of their environment, and when they have satisfied their scarce needs, they are at peace with nature and friendly to their fellow humans. With remarkable insight, Rousseau argues that the big change arrived with agriculture—a task so arduous that it required large numbers of people to be performed. "The first man who, having fenced in a piece of land, said *This is mine*, and found people naïve enough to believe him, that man was the true founder of civil society," states Rousseau.[4] This civilized man, unlike the savage, is never satisfied and always wants more goods, wealth, belongings, and even slaves. And though property can be acquired, it can also be lost. To quote writer Fernando Pessoa, "To possess is to lose."[5] The day before Rousseau's sudden death, the first of July 1778, a visitor told him, "Men are wicked," to which the philosopher replied, "Men are wicked, yes, but man is good."[6]

English philosopher Thomas Hobbes, who wrote that the life of a human, in its natural condition, is "solitary, poor, nasty brutish and short," opposed Rousseau's views; without the control of society, he argued, humans were violent and lived in a permanent state of war, everyone against everyone else.[7] Still, it would be an oversimplification to say that Hobbes believed humans were evil by nature; he was just pointing out that humans, in their natural state, were not meant to live in large, organized societies. But those who share Rousseau's views might legitimately ask, How can we achieve the best of our nature under modern social conditions? According to Wrangham, one way to solve what he calls the "Hobbes-Rousseau paradox" about human nature is again to consider that there are two types of aggressive behavior. Hobbes rightly recognized the potential of humans for highly proactive violence, while Rousseau noticed—rightly as well—the low frequency of reactive violence, especially among foragers.[8]

The "Neolithic Revolution"—a term coined by Australian archaeologist V. Gordon Childe in 1935—originated about ten thousand years ago in the Fertile Crescent.[9] It spread from there toward Europe some one and a half millennia later, and within just three or four thousand years the primordial way of life had been mostly abandoned. This change involved the biggest transformation experienced by the human species, not only in such aspects as demography and mobility, but in social structure. Nevertheless, it happened recently in the evolutionary history of the human lineage. A mere eight thousand years ago, all Europe's inhabitants were still foragers, and in principle they led the egalitarian, carefree, and kindhearted nomadic way of life suggested by Rousseau.

But despite Rousseau's idyllic depiction, neither the Mesolithic foragers nor the early farmers were always peaceful people. We have found archaeological evidence of violence and social hierarchy in sites from both periods. Although the evidence is probably more abundant in the farming communities—that were of course highly dependent on local resources and land for their survival—this raises the question of how the two communities interacted where they encountered one another.

The Mesolithic Age was coming at the end of the Last Glacial Maximum that took place between 26,500 and 19,000 years ago. Its chronology varies depending on the continental region, but it dates from about 11,500 to 5500 BCE. As the ice sheet disappeared, hunter-gatherer groups began to dwell on forested land and open tundra that had previously been uninhabitable. Hunting of species such as aurochs, red deer, and wild boar became a specialized task for Mesolithic hunters. This was accompanied by some technological innovation, most notably the general use of tiny flints in a wide variety of composite tools. Mesolithic human populations seem to have been small and highly mobile groups. Such mobility created a cultural homogeneity that might be reflected in some material objects, such as the ornaments made with atrophic red deer canines that can be found across European territories. It could also explain the remarkable genetic uniformity detected at the mitochondrial DNA level in the Mesolithic remains so far analyzed, which in most cases derive from the U4 and U5 lineages. (Different mitochondrial DNA lineages are labeled with letters by geneticists and can be thought of as the diverging branches of a family tree.)

Farming emerged in the Fertile Crescent, a Middle Eastern region that spans the Mediterranean Levant—extending to Egypt—and the regions between the Euphrates and Tigris Rivers, and was associated with the climatic improvement that took place after the Last Glacial Maximum. Local forager groups, known as Natufians, started collecting cereals whose seeds were somehow selected and planted again, which in turn made those groups less and less mobile—and increasingly dependent on the upcoming harvest.[10] Some stone structures from these Natufians can be glimpsed in areas of Israel such as the Golan Heights; they consist of simple rows that demarcate the perimeters of huts built on mud and cane where these pioneers lived.

These first settlements allowed the populations to become sedentary, which in turn enabled them to have more children. It must be remembered that foragers are compelled to move, sometimes across great distances, every day or every few days, and thus cannot have a new child until the previous one is able to endure long walks. Periodic harvests provided a stable food source with the possibility of generating surpluses, which could be stored

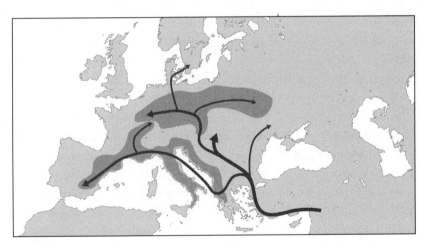

Figure 2.1
The spread of agriculture from the Middle East into Europe during the Neolithic. There are two main routes that we know of today that involved migrations from Anatolia: the Mediterranean and the Danubian (which reached central Europe). Gray areas mark the Impressa/Cardial culture distribution in the former, and the LBK (for *Linearbandkeramik* or Linear Pottery) culture in the latter. It has been suggested that the spread of Indo-European languages was associated with this migration. Modified from W. Haak, I. Lazaridis, N. Patterson, N. Rohland, S. Mallick, B. Llamas, G. Brandt, et al., "Massive Migration from the Steppe was a Source for Indo-European Languages in Europe," *Nature* 522 (2015): 207–211.

in silos for the next season. As a consequence, human diet shifted from an almost total dependence on animal prey along with gathered vegetables, fruits, roots, nuts, and mushrooms, to one primarily based on cereals, and afterward on domesticated animals and products derived from them such as milk. (This dietary shift also contributed to increased fertility, since cereal and milk could be made into a pap that allowed to toddlers to be weaned earlier, thus reestablishing women's fecundity.) Within a few hundreds of years, this new way of subsistence enabled a demographic increase that constantly required new arable lands. A wave of farmers penetrated Europe from Western Anatolia around eighty-five hundred years ago in what was undoubtedly a long and complex process; from there, some fanned out along the north coast of the Mediterranean while others went up the course of the Danube River into the central European plains.[11]

For decades, archaeologists have tried to grasp the nature of the interactions between local and incoming groups. The progress of these farmers across the continent can be traced with reasonable precision by artifacts such as pottery that foragers did not have; in contrast, domesticated animals are sometimes difficult to distinguish from their previous wild ancestors, at least during the first centuries or even millennia after domestication. At the population level, anthropologists have described physical differences between the two human groups, with the foragers being taller and more robustly built than the farmers.[12] The latter also suffered from a number of health conditions that have been attributed to interactions with livestock. Increased population density and close proximity to domesticated animals are thought to have resulted in larger numbers of potential hosts for infectious disease, enabling pathogens to spread among farmers; the same pathogens could not persist in low-density hunter-gatherer groups. A carbohydrate-based diet is thought to have contributed to the increasing prevalence of dental cavities, almost unknown to the foragers. Also, reliance on a few staple crops made farmers vulnerable to famine when those crops failed.[13] The relationship between an agricultural lifestyle and declining health can be illustrated by the drop in the adult height of European males from on average 179 centimeters during the Pleistocene to 150 centimeters during the Neolithic period.[14] Nevertheless, it is almost impossible to distinguish foragers from farmers by looking at the skeleton, not to mention understand their interactions, without relying on genetic analysis.

We can now use genetics to study the ancestry of the foragers and the first farmers across Europe, and uncover the dynamics of their interactions. The gist of these studies is that indigenous foragers did not adopt the farming strategy. Neolithization was a movement of people, not so much ideas: farming was spread by immigrants from Anatolia. But nor was it a complete replacement; we've been able to discern signs of the admixture of the languishing hunter-gatherers in the genomes of the incoming farmers. The first step in making this discovery was to understand the genetic ancestry of the Mesolithic hunter-gatherers.

In 2014, the complete genome of an eight-thousand-year-old Mesolithic male hunter found in the La Braña cave (León, Spain), high in the Cantabrian Mountains, was sequenced from the root's tip of a third molar by my research team.[15] The analysis of La Braña ancestry showed that he fell outside present-day European genetic diversity, but was actually somewhat close to modern northern Europeans—in particular Scandinavians—suggesting that these groups have retained more Mesolithic ancestry in their genomes than other modern groups. Interestingly, despite being separated by more than two thousand kilometers, the La Braña specimen showed an affinity to three Scandinavian late hunter-gatherers whose genomes were partially sequenced, which could be explained by the presence of a common genetic signature among European foragers.[16] This individual, however—as well as other Mesolithic foragers from western Europe sequenced later on—displayed a unique combination of physical traits. He had blue eyes but otherwise darker skin—the precise tone is difficult to determine—than all modern Europeans. Physically, early farmers from Anatolia were different from those foragers; they had brown eyes but fair skin and carried a whole range of mitochondrial DNA lineages, labeled by geneticists as N1a, T1, T2, J, K, H, HV, V, W, and X along with some of U lineage subvariants.[17] We can only guess how they regarded each other on first meeting.

Later genetic studies, including on several hundred Mesolithic and Neolithic individuals from other places across Europe, revealed the full picture of this transition. In distant places such as Great Britain, for instance, there seems to be an almost complete replacement of the local hunter-gatherers by farmers migrating from continental Europe; the latter showed substantial affinities with Iberian farmers whose ancestors followed the coastal Mediterranean route.[18] Only residual levels of local hunter-gatherer gene flow into the Neolithic genomes can be detected at sites in western Scotland. In contrast, genetic data from a large fishing community from eight thousand years ago at Lepenski Vir, located in the Iron Gates on the Danube shores (currently marking the border between Romania and Serbia), indicated that these people had farmer ancestry in variable ratios. Nine of these foragers showed mitochondrial DNA lineages that are typically found in early farmers (eight of them belonging to the K1 lineage and one to the H lineage). Two individuals, despite being buried as foragers in the settlement, were from a genetic point of view "farmers," which suggested that they came from abroad.[19] A person's degree of mobility can be ascertained from the analysis of the strontium and oxygen isotopes that accumulate in the dental enamel during tooth formation; this elucidates if someone spent their infancy somewhere other than where the skeleton was found because there are regional variations in the ratios of these isotopes in the environment. Indeed, when this technique was applied to the remains from the Iron Gates, it was possible to see that those individuals with a farmer ancestry profile did not grow up in the Danube area.

The case of the Iron Gates is not unique; for example, an individual with a genetic hunter-gatherer ancestry was found buried in a pioneer farming settlement from seventy-seven hundred years ago at Tiszaszőlős-Domaháza in Hungary.[20] Another individual with a similar chronology excavated at a site called Padina, also in Serbia, turned out to have a mixed farmer and forager ancestry, and was the first generation of an encounter between members of the two communities.[21] Elsewhere, such as at Early Neolithic sites in central Europe, some Balkan regions, and Iberia, evidence for early admixture is lacking, suggesting that at first contact the two communities perhaps avoided each other or their interactions were not necessarily that peaceful; maybe there were conflicts between them. In Iberia, for instance, the retrieval of an Early Neolithic genome from the Impressa/Cardial culture—named for the use of *cardium* shells to decorate pottery—dated to seventy-four hundred years ago, and coming from a site called Cova Bonica (literally, "Beautiful Cave" in Catalan) near Barcelona, revealed Anatolian ancestry without traces of local foragers similar to the ancestry found in early farming cultures from central Europe, including the prevalent LBK.[22] This suggests the existence of some cultural and reproductive boundaries between the forager and farmer communities, at least during the early stages of agricultural expansion.[23]

After this initial period of contact, we see an increase in the hunter-gatherer ancestry of the Middle Neolithic populations across most of continental Europe—corresponding, paradoxically, with the final disappearance of their way of life.[24] This trend seems the product of the final assimilation of local foragers into the incoming farmers' gene pool as they progressed inland throughout Europe. Curiously, it seems to have been driven by forager males and farmer females, at least in Middle Neolithic central Europe, Iberia, and the Chalcolithic Balkans.[25] As a result of the increase in hunter-gatherer ancestry in the Middle Neolithic, some typical Mesolithic Y chromosomes, such as I2, can be found in these farming settlements. Almost half the Y chromosomes in the Neolithic are G2a lineages, however, introduced by the newcomers. In contrast with other regions, genetic studies failed to detect such a resurgence of hunter-gatherer ancestry in the British Middle Neolithic, suggesting that the original Mesolithic populations were demographically tiny.[26]

When did the last European hunter-gatherers disappear? The last person with a fully Mesolithic-like genome—including the surprising combination of blue eyes and dark skin—was found at a Late Neolithic site in Poland and dated to sixty-three hundred years ago; they were buried alongside contemporaneous farmers.[27] It appears, paradoxically, that one of the last individuals with a forager-like genome was wielding a plow at the time of their death. Another person likely to have a fully Mesolithic-like genome (the current genetic data are incomplete) has been described in Cnoc Coig on the small island of Oronsay (Inner Hebrides, Scotland) and dated to just about six thousand years ago.[28] These are probably among the last representatives of an ancestry that was prevalent in Europe for more than ten thousand years.

One interesting observation from these studies is that the Neolithic European populations resulting from this encounter displayed an unbalanced ancestry, deriving most of their genomes from the Anatolian farmers and not from local Mesolithic Europeans. Although the foragers' ancestry is still discernible in the genomes of present-day Europeans, in whom it ranges from about 20 percent in southern Europe to 40 to 50 percent in Scandinavia and the Baltic region, no one today derives their entire ancestry from European foragers.[29] Therefore the final outcome of the Neolithic expansion was not only the disappearance of their way of life but also a dramatic reduction in hunter-gatherer ancestry across Europe.

That foragers were absorbed into the subsequent Neolithic gene pool doesn't mean that farmers consisted of peaceable groups that shunned violence. Archaeological evidence found in some early settlements indicates a need for protective defense. One of the oldest cities in the Fertile Crescent—perhaps the oldest—is Jericho, located on a hill that dominates the valley of the Jordan River, which flows from the Sea of Galilee to the Dead Sea in modern Israel. Successive cities were superimposed over it with time, forming an artificial mound (known as a tell in the Middle East). When archaeologists excavate, they often need to go through layers from ten different cities to reach the oldest archaeological levels. Jericho existed 11,500 years ago. By 10,000 years ago, it was already surrounded by walls and had at least one massive tower about nine meters high, built of undressed stones, from which the landscape could be dominated. As no known invasions took place during that period, the defensive nature of this tower has been debated, and some archaeologists suspect that it could have had a ritual significance.[30] Whatever the reality, it took an estimated 11,000 working days to build so was obviously intended as an imposing demonstration of power and domination. These first farmers then must have already had some degree of social hierarchy, at least in the larger communities.

Consistent with the archaeological evidence, there are signs of conflict too, especially among Late Neolithic groups in Europe, as bad harvests associated with deteriorating climate and the salinization of arable land began taking their toll. In 2006, a mass grave with twenty-six bodies of adult men and children was found during the construction of a road, twenty kilometers from Frankfurt.[31] The skeletons dated from 7,000 years ago and showed abundant signs of violence, from strong blows to the head to arrowheads in and around bones. More than half the individuals suffered from an intentional fracture of the leg bones—an action interpreted as a form of torture or ritual mutilation, perhaps to prevent the ghosts of the deceased from coming back to haunt the perpetrators of the massacre. Whatever the cause, this site is a palpable example of a farmer group that killed another, similar community in an outbreak of extreme violence. The conspicuous lack of young women among the dead also suggests that they were abducted by the rival group. At another Neolithic site in Halberstadt (Germany), the remains of nine adult individuals who died violently were discovered in 2013; at least seven of them were male.[32] Almost all of them displayed skull fractures caused by strong blows to the rear half of the head that were in all likelihood responsible for their death. The bodies had been thrown after a while into an abandoned silo, where they were piled up in awkward positions. Perhaps the most interesting evidence came from an isotopic analysis, which revealed that all of them were of foreign origin. We may never know what happened exactly, but it seems plausible that these farming groups were in fact small communities with strong family links that could either attack other communities or establish alliances with them. The instances of violence seem to be concentrated around the period starting 7,000 years ago, along with signs of environmental crisis. The tensions related to these processes would likely culminate in violent episodes as well as certain hierarchical trends.

For their part, the foragers certainly did not fulfill the Rousseauian ideal of peace-loving people. The publication in 2016 of the finding of twenty-seven bodies in Nataruk (thirty kilometers from Lake Turkana in Kenya), dated from around 9,500 to 10,500 years ago, informs us of a conflict between two bands of hunter-gatherers in an area that was at certain times rich in natural resources.[33] At least ten skeletons display traumatic injuries to the skull, and two more, judging from the general disposition of the bodies, died handcuffed. In this case, men, women, and children were among the victims of the massacre; one woman was in an advanced stage of pregnancy when she was killed, with her hands and feet bound. This is by no means an isolated case. During the construction of a railway track at the shore of a small lake near the Motala River in Sweden, the skulls of nine adults—four of them women—and one Mesolithic child, dated from about 8,000 years ago, were unearthed. The peculiarity of this site is that the skulls belonged to a secondary burial—that is, they have been exhumed from another grave—and had been deposited over a layer of rocks; at least two of them were impaled on a wooden stake.[34] Moreover, the skulls displayed obvious signs of violence in the form of trauma, and some showed evidence of healing from prior injuries, indicating previous episodes of violence. Due to the lack of other, similar examples from the European Mesolithic period, it is difficult to understand the significance of this site. But the public display of impaled skulls cannot be easily reconciled with ritual respect for ancestors, and it seems more like a warning of some kind. Genetic analysis of the remains indicated that at least two of the men were related, thus pointing to the possible extermination of a family by a rival clan.[35]

In conclusion, current genetic evidence suggests that the Mesolithic-Neolithic transition in Europe was a progressive and complex process, with regional variations in ecological conditions favoring agriculture and greater population densities, but nevertheless less traumatic and violent than subsequent, large-scale migration movements in Europe, as we will see below. In any case, the increase in population size driven by the transition to farming also meant that subsequent shifts in ancestry must have been more complex from a social standpoint and that they had to involve large numbers of people too.

The foraging-to-farming transition was not the last large migration to transform the European continent; thanks to new paleogenetic studies, we now know that the one associated with the steppe nomads resulted in an unexpectedly large-scale genetic change.[36] The final consequence of this new migration mirrored the previous one—in this case, a dramatic reduction of the previous European Neolithic genes all across the continent.

The Eurasian Steppe constitutes a vast ecosystem that stretches from Romania and Moldova to Ukraine, Russia, Kazakhstan, Mongolia, and Manchuria, and comprises grasslands and cold savannas. Russian writer Nikolai Gogol gives a romanticized description in his *Taras Bulba*: "Then all the South, all that region which now constitutes New Russia, even as far back as the Black Sea, was a green, virgin wilderness. No plough had ever passed over the immeasurable waves of wild growth; horses alone, hidden in it as in a forest, trod it down. . . . The whole surface resembled a golden-green ocean, over which were sprinkled millions of different flowers."[37]

Traditionally it was inhabited—and to some extent still is—by nomadic pastoralists based on horseback riding, as this ecosystem has too little rainfall to sustain farming. They relied mainly on horses, goats, and sheep—from which they obtained leather, wool, meat, milk, and cheese—to ensure the survival of their family clan. (The horse is still important to the economies of central Asian countries; in Kazakhstan, it is by far the most appreciated meat, to the point that the national team chose to bring its own stock of horsemeat to the London 2012 Olympic Games—much to the surprise of customs authorities.) Conditions during the cold winter would determine how many animals were to survive and thus how the clans would fare in their immediate future. The harshness of this environment meant that individual lives likely had less value than those of more sedentary societies and that kinship links as well as clan alliances were crucial for survival.

Nevertheless, a succession of favorable years across the steppes yielded a demographic surplus of horses, so people could move easily across its vast expanses.

Throughout history, highly mobile groups from the steppes periodically poured into the eastern, southern, or western Eurasian regions, with devastating effects on their local civilizations. In historical times, the Huns, Mongols, and Turks led such migrations, but at the end of the Neolithic, a rather unknown group of steppe nomads was key to a large-scale genomic reshuffling that affected all Europe.

About five thousand years ago, the farming communities at the end of the Neolithic period were living in crisis, as conditions worsened due to climatic change and soil salinization; also, a plague pandemic could have spread over trade routes and affected large settlements.[38] (As this pathogen had an Asian origin and was probably unknown in western Eurasia, no immunity had developed against it by local populations.) The archaeological record contains evidence of indiscriminate violence as resources decreased. This troubled Europe saw the arrival of yet another disturbance: the eruption of large-scale migrations of the steppe nomads that we call the Yamnaya (or Yamna).

David W. Anthony, author of the 2007 book *The Horse, the Wheel, and Language*, explains how Yamnaya culture emerged from the Pontic steppe, a vast geographic area encompassing the lands between the Don and the Volga Rivers as well as a large expanse north of the Caucasus and the Caspian Sea.[39] It was characterized by highly mobile pastoralist groups that based their economy on horses, cows, and to a lesser extent, sheep; the Yamnaya might also have hunted and fished, and even practiced occasional farming on river terraces. But most significant, they had horses— and perhaps they even were responsible for having domesticated them, as the earliest evidence of horse harnessing and corralling is found in the fifty-five-hundred-year-old central Asian culture called Botai—and held an additional military advantage: wagons suitable for transport and warfare.[40] Petroglyphs depicting horses and wheeled carts have been found at numerous Central Asian sites, from Kazakhstan to southern Siberia, and some have been dated to the critical period of the Yamnaya migration.[41] This means that they could travel faster and farther than any of their predecessors. (It is said that the much later Huns could eat and even sleep while riding.) Horses undoubtedly represented a military and psychological advantage, as the heavily outnumbered Spaniards realized in their conquest of the vast Aztec and Inca Empires. What facilitated the emergence and spread of the Yamnaya remains unknown, but it seems to correlate with a period of optimal climatic conditions for the steppe environment that lasted from 3600 to 3300 BCE, guaranteeing a stable enough exploitation of natural resources to sustain large populations. Additionally, the fourth millennium BCE was a period of technological innovation in riding that further enhanced human mobility.

Another distinctive trait of the Yamnaya pertains to their burials. Unlike previous collective burials, they buried their dead in individual tombs under barrows, with the body in a crouched position. Some of them, presumably the most important chieftains, were buried under large tumuli (called *kurgans*), usually accompanied by grave goods and sacrificial offerings— mainly animals—that were placed alongside the body. This trend and the existence of large kurgans that likely took weeks to build by hundreds of laborers suggests a formidable social hierarchy, unprecedented in the European continent.

The Yamnaya turmoil precedes the emergence of a widespread archaeological horizon around 2900 BCE, covering most of Scandinavia along with central and eastern Europe—from the Volga to the Rhine Rivers— and known as Corded Ware. The name derives from a type of pottery that is decorated with cord-like impressions across its surface. The horizon is characterized by other elements as well, such as the widely present battle axes, which served as a symbol of male dominance. The lack of large settlements, in contrast with what has been observed in previous periods, suggests a nomadic way of life in which they probably assembled light-framework dwellings that were transported by animals, while the burning of large forest extensions to create pastures indicates that they were more closely associated with pastoralism than any local farming development.[42] The existence of single inhumations under barrows, considered a fundamental identifying sign, links the Corded Ware to the Yamnaya people, according to the Danish Bronze Age expert Kristian Kristiansen and colleagues.[43]

The Yamnaya were genetically distinct from the Late Neolithic Europeans. Their ancestry derived in turn from a blend of two still-different predecessors, both Mesolithic: the so-called Eastern Hunter-Gatherers, present from the Baltic shores to the Urals before the arrival of farmers, and

Caucasus Hunter-Gatherers, inhabiting the eponymous region. Due to the differences between Europe and the steppe, the ancestry shift associated with the arrival of the Yamnaya can be examined quite precisely. Genetic analyses of the skeletal remains found in the Corded Ware archaeological context have yielded a nearly 79 percent proportion of Yamnaya-like ancestry (the rest being identical to previous Late Neolithic substrata).[44] This large component of steppe nomad ancestry explains why in general genetic terms, the Corded Ware people appear to be closer to the Yamnaya than to any other previous local group, despite being separated from the Pontic region by over twenty-six hundred kilometers.

The number of Yamnaya immigrants must have been significant because contrary to what we've seen in the Mesolithic-Neolithic transition, the Late Neolithic populations in Europe were already substantial. But central Europe was not the last stop for these genetic changes: the arrival of the

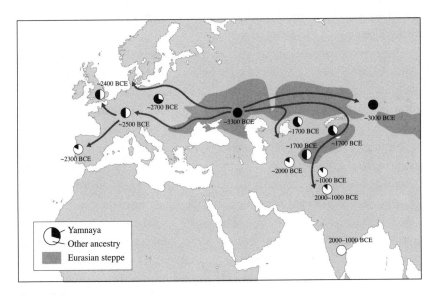

Figure 2.2
The genetic impact of the steppe ancestry brought by the Yamnaya in both Europe and South Asia. The percentage of Yamnaya ancestry (*black*) versus other genomic ancestries is represented in the circles. The arrows mark potential population movements to western Europe and South Asia (see chapter 4), and also a reflux to the east. *Source*: V. M. Narasimhan, N. J. Patterson, P. Moorjani, N. Rohland, R. Bernardos, S. Mallick, I. Lazaridis, et al., "The Genomic Formation of South and Central Asia," *Science* 365 (2019): eeat7487.

steppe ancestry in different corners of Europe can be detected hundreds of years later, reaching as far as the British Isles and Iberia.

These steppe nomads carried essentially two distinct Y chromosome lineages: R1b—or more specifically, the sub-subgroup R1b1a1a2—and R1a—specifically R1a1a. Neither lineage was at all present in western Europe prior to the arrival of the steppe nomads. (A different branch of the R1b lineage has only been found in two individuals from Italy and Iberia, dated to fourteen and seven thousand years, respectively.) After their arrival, both Y chromosome lineages increased until they become prevalent in modern Europeans. (R1b goes from 41 percent of all Y chromosomes in Germany to 83 percent in Ireland, and R1a reaches 50 percent of male lineages in Scandinavia and northern Europe.) The previous Neolithic Y chromosomes—predominantly G2a and I2—almost disappeared from the continent, which means that most men from that period were literally unable to continue their paternal line. By contrast, the mitochondrial genetic composition stays relatively unchanged, and we continue to find the whole range of lineages present in the previous period.

The newcomers also likely provided novel genetic variants, some of which became prevalent after being selected in the resulting admixed populations. Probably the best example is a mutation in the lactase gene that allows its carriers to digest milk during adulthood. This mutation has been anecdotally described in Bronze Age continental Europe and was seemingly absent from the continent before forty-two hundred years ago. Moreover, it has been strongly selected only in the last twenty-five hundred years until becoming literally fixed in such places as Scandinavia and the British Isles.[45] (It is less prevalent in the south of Europe; I myself have only one of a possible two copies of this mutation, which probably derives from my English grandfather.) Probably it represented a large advantage to survival in those regions where harvests could fail due to unfavorable climatic circumstances and the possibility of being fed with animal milk turned out to be critical for survival. Interestingly, the genetic differences were not restricted to the lactase gene; the newcomers were apparently taller than the local European farmers.[46] This physical trait can be ascertained from both their genomes (more than seven hundred genetic variants associated with height are known, so it is possible to detect from genetic evidence alone) and their skeletons; the resulting information could add yet another new dimension to this migration into Europe.[47]

As might be imagined, the arrival of the newcomers was not without violence. Genetic analysis of fifteen skeletons found in a mass grave in Koszyce (Poland) containing males, females, and children who were brutally executed by blows to the head dates the incident to the critical contact period, around forty-eight to forty-nine hundred years ago.[48] Genetics uncovered half a dozen family links as well as the surprising findings that in most cases, adult males were absent and the dead seemed to be buried by people aware of their family affiliations, placing two brothers side by side and a mother holding her child. A plausible scenario is that the adult men were away—or had fled—when the massacre took place and they arrived later on to bury their dead. As the people from this mass grave show no steppe ancestry, and the place and period correspond to the territorial expansion of the Corded Ware groups, it is possible that we are witnessing the result of a violent intergroup conflict associated with the observed genetic turnover.

Also, the retrieval of the *Yersinia pestis* pathogen adapted to infect fleas that acted as a plague vectors—that is, in its most virulent form—from some skeletons of that period with steppe ancestry suggests that locals may have been stricken by this pandemic.[49] (It is currently unclear how the plague strains found in some Late Neolithic individuals had infected humans.)[50]

The Yamnaya invasion represents not only a large-scale genetic disruption but also an indirect turning point in prehistoric Europe, marked by the subsequent formation of social elites. Later archaeological horizons and cultures show, with few exceptions, clear signs of inequality, with people being inhumed in "princely" tombs, sometimes with lavish grave goods, alongside others entombed without a thing. The story of the steppe ancestry is not yet over, as we can trace its spreading from central Europe to the western corners of the continent in great detail, albeit under different archaeological contexts and temporal horizons.

The Bell Beaker complex is one of the best-known and widely studied archaeological horizons of European prehistory. The Bell Beaker complex seems to have emerged during the Chalcolithic period (ca. 2900 BCE) around the Tagus estuary (Portugal), and quickly spread throughout large regions of western, northern, and central Europe, ranging from the Vistula River to the Atlantic coast, and from Iberia and Sicily to Scotland. For a prehistoric culture, this represents an unprecedented geographic scale that is only comparable in size to the modern European Union. The whole Bell Beaker horizon history lasted for about a thousand years, with some geographic variations along with a longer persistence in northern Europe and the British Isles.[51]

The central objects of the Beaker assemblages are the distinctive bell-shaped beakers related to the drinking of beverages of ritualistic significance—especially beer—as well as archery equipment and copper daggers symbolically associated with another sphere of life: war and hunting. These objects had significance for their users because they were often interred with them in single burials where the skeleton was deposited lying on one side in a flexed position. From an archaeological point of view, the Beaker horizon is remarkably uniform—especially in its initial stages—although it subsequently diversified under the influence of regional Neolithic traditions. The conspicuous spread of this complex is associated with a period of prosperity that from an economic point of view seems to be related to the expansion of copper metallurgy. After the decline of the Beaker complex, the Early Bronze Age was dominated by new cultural developments—including, as its name indicates, bronze metallurgy—and the emergence of fortified settlements with an increase in social stratification and violent conflicts.

Recent paleogenomic studies have shown that the original Bell Beaker complex spread into central Europe involved no discernible movement of people—or genes, although a subsequent reflux from central Europe into such places as Great Britain, Iberia, and northern Italy shows a spread of the steppe ancestry, then prevalent in the central areas of the continent.[52] Genetic surveys have detected an almost complete substitution (over 90 percent) of the British Isles' Neolithic ancestry—in parallel with the Y chromosome gene pool—by four thousand years ago with the arrival of the Beaker artifacts, probably from the Netherlands or Germany. Again, the distinctive steppe ancestry is what enables geneticists to detect this shift in ancestry, which is probably unprecedented in ancient migrations in the sense that almost no overlap between the previous Neolithic ancestry and the steppe ancestry-like newcomers can be detected. The former—the ancestry of those who built Stonehenge—is simply gone, both at the autosomal and Y chromosome level. From then on, the prevalent paternal lineage in Britain is the Yamnaya-like R1b. In this case, an almost complete replacement can only be explained if men and women alike migrated. And indeed, this seems to be the case, because the maternal counterpart of the Y chromosome—the mitochondrial DNA—shows the arrival of lineages that

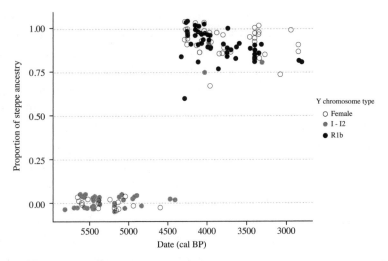

Figure 2.3
The spread of steppe ancestry into Great Britain during the Bronze Age. This migration is associated with the arrival of the Bell Beaker complex and marks an abrupt substitution, until then unparalleled in Europe, of the previous Late Neolithic population. *cal BP*: "calibrated Before Present" refers to a method used to correct for carbon-14 fluctuations over time; it yields more precise dating than raw radiocarbon dates. *Source*: I. Olalde, S. Brace, M. E. Allentoft, I. Armit, K. Kristiansen, T. Booth, N. Rohland, et al., "The Beaker Phenomenon and the Genomic Transformation of Northwest Europe," *Nature* 555 (2018): 190–196.

are found among the Yamnaya too, notably those named I and U4. Some archaeologists have suggested that the skeletons of the local British may not have been adequately sampled. But the question is that in subsequent periods, we see no resurgence of the previous local ancestry, as observed during the Middle Neolithic with the forager ancestry. During the Late Bronze–Early Iron Ages, however, there was yet another migration that could be traced back to Alpine regions in continental Europe; this subsequent influx into the British Isles could have spread early Celtic languages.[53]

During the Bronze Age period between 2400 and 2000 BCE, almost half the ancestors of the Iberian people can be traced to incoming groups from central Europe, mirrored by an almost complete substitution of the Neolithic Y chromosomes by the newcomers' R1b.[54] Of course we don't know the details of this process, but we can assume it was to some extent a dramatic period for the locals. Strong social biases both in the net reproduction and offspring survival rates could help explain the results observed after the overlapping period. This scenario has been challenged by some Spanish archaeologists, who did not perceive it from the archaeological record, beyond a change in the location of the settlements that moved to more defensive positions during the Bronze Age. In any case, genetics has the potential to launch new hypotheses that can in turn be tested by other disciplines. With this information at hand, we may need to take a fresh look at the archaeological record.

An interesting question raised by unbalanced transitions such as the one observed in Iberia is, What were the social mechanisms driving them? And how did these events affect such cultural tools as language?

Language is both a primary tool for social transformation and one of the strongest signs of collective identity. It clearly establishes a specific context that constrains both the life of its speakers and their particular view of the world, and in many ways acts also as a genetic boundary by being a cultural barrier and thereby sometimes a barrier to mating. Moreover, language is a mechanism for social organization as well as political control. Modern sociolinguistic theory has established strong links existing between language use and inequality, mainly from work done on colonial empires. Now that we can reconstruct past migrations and population shifts, we might be able to tackle the spread of languages and language families in a new light.

One of the best known of such language families is the Indo-European. Since the work of Sir William Jones in 1786, philologists have known that most languages spoken today in Europe—and others such as Sanskrit in India—have a common origin and can be included in a single language family. The name was coined in 1813 by Thomas Young, an English physicist and linguist who competed with French scholar Jean-François Champollion to be the first to decipher Egyptian hieroglyphs. It includes the Latin, Germanic and Baltic-Slavic languages, Celtic, Greek, Armenian, Tocharian (spoken in the west of China until perhaps the ninth century CE) and the Anatolian languages (which included Hittite, from which texts dating from 1650 BCE are known), among others.

How such a large and diverse set of related languages spread was a subject for debate a few decades ago. In 1987, the archaeologist Colin Renfrew suggested that the phenomenon could be explained by a demographic transition that likely had unique continent-wide impact: the spread of the Neolithic across Europe, starting about eight thousand years ago.[55] The foundational position of Hittite among modern Indo-European languages could tie in with the origin of farming in Anatolia. This was certainly an attractive hypothesis because of the expected demographic and social transformation associated with farming, and was favored by archaeologists and geneticists—less so by linguists, interestingly—for several decades. Some researchers, however, pointed out the existence of common words among the Indo-European languages that referred to equine objects, such as harness, wheel, and wagon axles that were clearly post-Neolithic.[56]

Yet the paleogenomic studies undertaken in Europe over the past few years have revealed the scale and genetic impact of the steppe-nomad migration at the end of the Neolithic and subsequent periods, while noting that Basque ancestry is not Mesolithic; if anything, modern Basques are most similar to the Iberian Iron Age people and clearly have as much steppe ancestry as other Iberians.[57] Also, it is in the Basque country that the typical Y chromosome steppe lineage, R1b, is most prevalent—up to 80 percent in males. Genetic researchers drew a clear inference from this: if changes occurred much later than the Mesolithic-Neolithic transition, a Neolithic origin for the Indo-European languages is less plausible.

Even the origin of the intriguing Tocharian languages in the Tarim Basin in China can be explained as a result of a reflux movement of steppe ancestry from west to east. In a recent study of ten individuals from the Afanasievo culture, excavated at the Shirenzigou site in northeastern Xinjiang (China) and dated to twenty-two hundred years ago, it was revealed that they carried a Yamnaya-like ancestry in 20 to 80 percent ratios (the rest being of eastern Eurasian ancestry).[58] Consequently, four out of six males also carried the common R1b Y chromosome lineage.

That the steppe nomads penetrated the continent across the eastern European plain conflicts with the fact that Anatolian languages—which include Hittite, along with Luwian, Lyiciand, Pisidian, and other extinct languages—are considered the oldest branch of the Indo-European family. This foundational language might have arrived in Anatolia through the southern Caucasus from having a common origin with other early Indo-European languages in Central Asia. In a genetic study published in 2018, five Anatolian samples contemporaneous with the Hittite Empire were analyzed, and none of them showed signs of steppe ancestry. The researchers interpreted this to mean that the Anatolian languages—or at least Hittite—perhaps did not arrive by migration from the steppe.[59] Anthony stated, though, that the individuals probably did not belong to the royal Hittites and might actually have been Hattians (people from the land of Hatti in central Anatolia), who spoke a non-Indo-European language before being forced to adopt Hittite between 2000 and 1700 BCE. It could be that the imposition of Indo-European languages in Anatolia was related to a dominant elite whose steppe ancestry would be subsequently diluted. Alternatively, it could have been an example of cultural diffusion affecting the oldest branches of the family. But this is not the only example of the complex interactions between genes and language in prehistory.

Genetic research on the Iberian Peninsula pointed out some difficulties in generalizing a correlation between large-scale migration and language substitution. Besides the dramatic population replacement observed in the Bronze Age, there is evidence of subsequent gene flow from central Europe during the Iron Age (with about 20 percent of additional steppe ancestry arriving after the Bronze Age event).[60] Yet at the time of the arrival of the Romans hundreds of years later, the Mediterranean half of Iberia—as well as the "Vascones" or ancient Basques—spoke non-Indo-European languages, while the western half spoke Celtic languages—thus Indo-European like all the languages of the Celtic family. This heterogeneity is perplexing, but it also sheds light on the complexities of interactions between incomers and locals on languages. In short, population change is a necessary but not sufficient criterion for language change.

It may come as a surprise that ancient DNA can shed light on the linguistic expansions of the past. While genetics obviously cannot attribute a specific language to certain people, whatever their ancestry, it can definitely establish the demographic and genetic magnitude of the migrations that took place, and as we have seen, the social transformations associated with these migrations, likely including language replacement. In this sense, the spread of languages becomes another collateral effect of inequality, at times speaking to us from the distant past.

The ancient migrations that have shaped the diversity of modern human populations and can now be documented by ancient genomics also set the stage for further exploration of how these transformative demographic events influenced different aspects of human societies. Reich's book had a positive message because these repeated mixtures constitute a scientific argument against "pure" populations and racial classifications.[61] What I reveal in the following chapters, however, like an ominous epilogue, is that the story of these past migrations is one of inequality.

3   Archaeology of Inequality

Kings made tombs more splendid than houses of the living.
—J. R. R. Tolkien, *The Two Towers* (*The Lord of the Rings* trilogy)

November 26, 1922, marks what is arguably the most famous discovery in the history of archaeology. On that day, the British Egyptologist Howard Carter made a small hole through which he could insert a candle in the sealed doorway of Tutankhamun's burial chamber and thus lit the interior. As his eyes slowly adapted to the darkness, he was able to make out a chamber that had not been disturbed for over three thousand years. An astounded Carter described the moment: "Details of the room within emerged slowly from the mist, strange animals, statues, and gold—everywhere the glint of gold. For the moment—an eternity it must have seemed to the others standing by—I was struck dumb with amazement, and when Lord Carnarvon, unable to stand the suspense any longer, inquired anxiously, 'Can you see anything?' it was all I could do to get out the words, 'Yes, wonderful things.'"[1]

Tutankhamun was just an obscure pharaoh during his lifetime, and there is evidence that he was hastily buried; the second of the three nested coffins seems to have originally belonged to someone else, maybe his half-sister Meritaten—and yet the inner coffin is made of solid gold and weighed 110.4 kilograms (on the current rising gold market, this piece alone would fetch more than half a million dollars). One can barely imagine how impressive the burials of such powerful leaders as Kufhu, Tuthmose III, or Rameses II must have been; alas, they were all looted in antiquity.

Besides the famous discovery of Tutankhamun's tomb, there is widespread evidence that outstandingly rich people existed in the past. If anything, this perception may likely increase in the future, because a long list of graves of famous kings and conquerors still remain to be discovered, among them Genghis Khan's and Alaric's. Qin Shi Huang, first Chinese emperor, was buried with an army of more than eight thousand terra-cotta soldiers in the year 201 BCE, and his body is supposed to rest amid fabulous wealth in an underground palace that eludes excavation as it is immersed in a lake of mercury. An imposing tomb discovered inside the Kasta tumulus in Greece in 2014 has been attributed not to Alexander the Great—who is known to be buried at Alexandria, a city founded by him, as the name suggests—but instead to someone else from the Macedonian royal family or nobility, obviously much less important than the conqueror of Persia. Even such an obscure king as Antiochus I, from the rather minor kingdom of Commagene, was able to build an enormous tumulus flanked by gigantic statues atop a 2,134-meters-high mountain in Turkey called Nemrut Dagi (his tomb is supposed to be underneath it, though this has yet to be ascertained).

Contrary to popular belief and cinematic glorification—not to mention romantic adventures of the sort described in *Gods, Graves and Scholars* by C. W. Ceram—most archaeologists would say that the search for spectacular treasures isn't their main research objective but an attempt to understand the daily life of past civilizations.[2] Still, both extremes—the fabulous wealth of kings and the hardscrabble existence of common people—contribute to an understanding of what can be argued is one of the main goals of archaeology: to document and study the evolution of inequality in ancient societies. But this also involves the question of how to recognize and quantify it. As we have seen, one of the most obvious approaches would be through the assessment of differential goods deposited in contemporaneous graves. One doesn't need to be a powerful pharaoh or boast ostentatious funeral displays; some people are found buried with a knife or pot near contemporaneous peers who had nothing at all with them for their trip to the afterlife. Richly furnished graves, however, may not simply be evidence of social differentiation; rather, they may be an attempt to demonstrate the importance and distinction of a family in relationship to other kindreds—a social importance that may not exist in reality. Sometimes there isn't even any indication of potential social differences, but just different locations within a cemetery. Moreover, social stratification can be based on wealth, as evidenced by the dead, but can also be based on personal prestige and power. In conclusion, it isn't always possible to assess social differences by comparing graves with goods to those without them.

Some archaeologists have attempted to apply economic principles to examine social differences at specific sites and, crucially, compare the data from different places. A study published in *Nature* in 2017 and led by Samuel Bowles from the Santa Fe Institute tried to address this question by applying the Gini coefficient across a large number of sites from the archaeological record, both in the Old World and the Americas (starting much later in the latter).[3] The list of sites included paradigmatic cities such as Catalhöyük in Turkey, Pompeii in Italy, and Teotihuacan in Mexico; the authors measured the dimensions of houses as estimated indicators of wealth.

Among modern hunter-gatherers, the Gini coefficient is low—around 17 (on a scale of 0 to 100). This is not surprising as few objects can be carried in nomadic societies, and consequently, personal qualities such as the ability to hunt count for more. (Even modern hunter-gatherers walk remarkably long distances daily depending on food resources.) For instance, some objects of high symbolic value for the Paleolithic hunter-gatherers, such as the famous Venus figurines carved mainly in ivory, clay, or stone, were small enough to be portable. Interestingly, they were not deposited in burials; they've always been found in habitational layers. This does not mean that some people didn't have a higher social status; material culture was probably so poor—or so different from our perceptions of status—that it is difficult to grasp social differences among past hunter-gatherers. In 2008, for example, a Natufian individual burial—the Natufians are precursors of the first farmers in the Near East—dated to 10,800–9500 BCE, was found in the Hilazon Tachtit cave (in Galilee, Israel).[4] It corresponded to a diminutive adult female with a deformed spine and pelvis—possibly a shaman of the group. Within the burial pit, she was surrounded by bones and remains that seemed to have a magical significance, including seventy-one tortoise shells, a human foot, the tail of a cow, the wing bone of an eagle, two marten skulls, the leg of a pig, and a leopard's pelvis. Some suggest that a ritualistic feast took place in her honor prior to her burial, as some shells were burned—indicating that the meat was roasted—and some bones had cut marks. We might think that these are not especially significant offerings, but clearly they had an important meaning for the rest of the community because they were carefully arranged around the body and also because these objects are conspicuously absent in a collective burial pit of twenty-eight other bodies within the same cave. In any case, evidence of an organized religion and a person likely in charge of connecting the living to the dead indicates that some social inequality, perhaps associated with the emergence of new economic practices such as cereal harvesting, might be present as early as twelve thousand years ago,

But in the ancient farming societies under study, the Gini coefficients are estimated to have been around 35 to 46; interestingly, the real measurements were lower than those obtained from contemporaneous chronicles.[5] For instance, among the ruins of Babylonia, researchers estimated a coefficient of 40, yet an estimate based on information from the Babylonian chronicles resulted in a higher coefficient of 46. The ancient accounts likely overemphasized the size of the largest houses in admiration. This is not unlike what happens when we return from a trip: we sometimes tend to exaggerate the things that we've seen. (Every tourist is a bit like Marco Polo, whose tome on his voyages came to be known informally as "the book of the million lies.")

Nevertheless, the most remarkable differences come from the comparison of the societies of the Old World and those of the Americas, with the latter being much more equal in the Gini coefficient, despite being highly hierarchical in some cases such as the mighty Aztec Empire. Researchers conclude that the root of these differences could be ecological since there were more and larger animals to be domesticated in Eurasia—such as cows, horses, pigs, sheep, and goats—than in the Americas—with only dogs and turkeys—and this trait alone created a differential system of accumulated wealth. At the Aztec capital, Tenochtitlán, for instance, houses had highly standardized dimensions and were all quite similar. Aztec society, even with its horrific human sacrifices, was at the time of the Spanish conquest more egalitarian than Mexico two hundred years later, when the European elite had created the encomiendas system, under which the indigenous population worked in semislavery. Within a few generations, the concentration of wealth had almost doubled in the colonial New World, with a consequent increase in inequality.[6]

When did these differences between the Old and New Worlds emerge? Early farming societies had the possibility of generating—and storing—food surpluses, creating potential scenarios for differences in population size along with a certain degree of inter- and intrasettlement inequality. A recent application of the Gini coefficient to ninety sites from the Near East and Europe showed a remarkable increase of inequality thousands of years after the advent of agriculture—a finding that would indicate it was not farming per se that created unequal societies.[7] According to the authors, at some point some farmers were able to maintain specialized plow oxen that could cultivate ten times more land than other farmers, thereby transforming the economy toward a higher value of land in detriment of human labor. This emerging inequality at the end of the Neolithic could explain a remarkable example of wealth dating from that period: the Varna burial. This burial was found in a Copper Age cemetery in modern Bulgaria and is dated to 4560–4450 BCE; it contained more gold than the rest of the world possessed at that time.[8] It contained an adult male—likely a chieftain or king of some sort—who was buried holding a gold war mace; curiously he also had a gold penis sheath of unknown meaning. Still, such findings are exceptional, and there is a general consensus that Neolithic societies were more egalitarian than later ones.

Figure 3.1
A re-creation of the skeleton and funerary artifacts found in the Neolithic necropolis of Varna in present-day Bulgaria. Dated to about sixty-five hundred years ago, the burial is the oldest gold treasure in the world. It is currently displayed at the Varna Museum of Archaeology (Bulgaria). Image from Wikimedia Commons (Creative Commons Attribution—ShareAlike 4.0 International).

Inequality clearly increased with the arrival of metals, which partly allowed, from 3000 to 2000 BCE onward—as we saw in the previous chapter—the appearance and development of a social organization based on the emergence of elites. Once the initial power structure was established, it attempted to perpetuate itself dynastically by increasing social control and building up familial alliances with other chiefs. Control mechanisms often involved violence. The possibility of using horses—and to lesser extent, camels—as instruments of war determined the success of conquests that would alter the pattern of settlements across Eurasia at the end of the Neolithic. This would at least partially explain how thirty empires or large states that emerged between 3000 and 600 BCE were all found in the Old World, where these animals roamed.

Consequently, tombs with signs of wealth became more abundant in the archaeological record, such as the famous Amesbury Archer, found three miles southeast of Stonehenge in 2002 (near today's Salisbury) and dated to 2300 BCE.[9] This grave includes more artifacts than any other Bronze Age British burial; besides numerous arrowheads, three copper knives, four boar's tusks, two stone wrist guards that protected users from their bowstrings, and five pots that conformed to the Bell Beaker tradition, there were two gold hair ornaments—the earliest pieces made of this metal ever found in the British Isles. The arrival of the Bell Beaker complex to the British Isles is associated with an almost complete replacement of the prior local population and subsequent emergence of social elites. The Amesbury Archer must be considered in the context of the spread of metalwork and supraregional exchange networks in a process that archaeologists sometimes call "Bronzization."[10]

There is evidence of early kingdoms from around that time with the ascendance of the pharaohs of Egypt in the eastern Mediterranean as well as the emergence of the Sumer civilization between the Tigris and Euphrates Rivers. Yet even in such remote areas as the southeast of the Iberian Peninsula we find, with the El Argar society around 2200 BCE, the first appearance of large palatial structures. These include the courtroom found at La Almoloya in Murcia, with a capacity of sixty-four people seated on benches along the walls. Moreover, archaeologists have confirmed the existence of high-status people there; in 2014, two such individuals were found buried within a large storage vessel in a privileged location next to the main wall of the palatial hall. The tomb contained a man and woman along with thirty objects made of silver and gold as well as semiprecious stones. Among the most impressive objects was a silver diadem that encircled the head of the woman (four identical diadems were found in early twentieth-century excavations at other El Argar sites) as well as gold and silver ear dilators. Aside from this princely burial, few other graves at the same site display a significant number of funerary goods—and many none at all—indicating the presence of substantial social differences.[11]

The increase in inequality in Bronze Age societies is reflected as well in the appearance of settlements of a clearly defensive character. During the previous Neolithic period, villages were established on the plains, surrounded by crop fields and near streams or rivers. Many of these settlements were abandoned during the transition to the metal ages, while others appeared on fortified and difficult-to-access hills. At another El Argar site, La Bastida, excavators have uncovered two- to three-meter-thick walls reinforced by solid towers measuring four meters on each side and that would have stood up to six or seven meters high.[12] Aside from the potential regional conflicts suggested by the presence of such massive walls, the efforts required to build such formidable constructions must have required significant social differentiation.

The rise in inequality during this period, both in the Middle East and, as we have seen, parts of western Europe, seems to be partly influenced by an increase in population density. (Incidentally, population size was always larger in the Old World than in the Americas, notwithstanding the large size of such cities as Tenochtitlán.) This correlation is likely related to a growing complexity in modes of subsistence, trading networks, and political organization associated with population growth. Although the highest Gini coefficients for past societies determined by the Santa Fe Institute were similar to those found in some present-day European countries (for instance, with values of around 60 in Pompeii and Kahun, an Egyptian settlement from the twelfth dynasty), they remained below the values for the most unequal modern societies such as China and the United States (with Gini coefficients of 73 and 85, respectively), which obviously have larger populations.[13] From a historical perspective this would suggest that an increase in population size brings higher inequality—an issue explored by Piketty in recent times, but that likely has parallels in Bronze Age populations.

Still, the Gini coefficient cannot always be applied since some settlements have grown with time over the destruction of previous ones, piled one atop another like the layers of a cake. Many ancient sites could not possibly be studied in detail; for instance, at Hisarlik—the old Troy—at least ten cities arose atop their predecessors in just two thousand years, making them quite difficult to disentangle.[14] In addition to this limitation, whether the Gini coefficient can be transferred between different cultural, geographic, and ecological environments to make direct comparisons has also been a subject of debate since such factors can influence their inhabitants differently. For example, a settlement established in a jagged terrain would favor smaller, more vertical houses than one extending over a vast plain.

The economic interpretation of past settlements has received some criticism from among the archaeological community; some argue that the quality and solidity of the building materials can be as important as the size of the houses. In our modern cities, we're all aware that location—for instance, close to the city center—is usually more important than size. Finally, the ostentatious wealth—opulent furniture, wall paintings, mosaics, and so on—that can still be found in some excavated houses such as at Pompeii—should be taken into consideration too, though such features aren't usually well preserved.

One way around these limitations might be to compare the Gini coefficients with the so-called health inequality of each population since buried human remains are sometimes better preserved than buildings. There are several skeletal indicators (dental cavities, arthrosis, traumas, vitamin deficiencies, etc.) that can reflect the health status of the population in each period. The frequencies of these pathological markers are in general higher during periods of higher inequality. For example, the 2006–2013 excavation of nonelite cemeteries such as North Tombs Cemeteries at Amarna demonstrated deaths at an early age—mainly of children, teenagers, and young adults—widespread dietary deficiencies, and indications of hard labor, suggesting the poor state of health and substandard working conditions for most of this urban community. For instance, 16 percent of all children under fifteen displayed spine injuries of the sort associated with carrying heavy loads; none of them had any grave goods, and sometimes were buried together with several others, with scant regard for the disposition of the bodies—a grim image that contrasts with the glamorous depictions of the pharaoh's family in the Amarna style.[15]

An additional indicator would be evidence of a high infant mortality rate, although the preservation of children's skeletal remains is invariably more difficult than that of adult bones due to differential conservation processes, and this could represent an insurmountable bias in the results. Changes in health status can be used to ascertain cultural and ancestral transitions too. In this sense, probably the most striking change observed is between hunter-gatherers and the first farmers in Europe. The latter not only show signs of poorer health—such as cavities, almost unknown by the former—but also higher infant mortality rates and even lower stature than previous hunter-gatherers.

Correlated with this information, recent developments in the stable isotope analysis of carbon and nitrogen ratios in bone collagen can provide information on nutritional status and mobility patterns associated with specific individuals. For instance, the analysis of a high-status burial in Helmsdorf, Germany, related to the Únêtice culture, showed that this person had a higher protein intake than other contemporaneous peers, suggesting as well that diet can be as much an indicator of social status as it is in today's societies.[16]

Key to understanding the social panorama of the past is that ancient cemeteries can provide not only potential indicators of inequality in the form of grave goods and even differential health status but also genetic material preserved within human remains. The information retrieved from their DNA can be used, for the first time, to correlate ancestry with social power in each period. Furthermore, a crucial aspect of the accumulation of power is the possibility of bequeathing wealth to biological relatives— something that can be tested as well via the interface between genetics and archaeology, which enables us to reveal family links. (Unfortunately, neither Tutankhamun nor other pharaohs' mummies have been subjected to solid paleogenetic analyses yet.)

Like funerary goods, a privileged resting place could serve as a status marker too. Around sixty-five hundred years ago, the phenomenon of building large funerary stone structures—known as megalithic tombs—emerged, mainly across Europe's Atlantic seaboard, and culminated in the great passage tomb complexes such as Newgrange in Boyne Valley (Ireland), which has a mound eighty-five meters in diameter and thirteen meters high. The origins and meaning of these monuments, which required a heavy investment in labor, have been debated for more than a century, as has the social organization of the farming communities that built them. The genetic analysis of two-dozen individuals found in various megalithic tombs from

Scandinavia to Orkney Island and Ireland yielded some interesting social clues.[17] In some places, notably the British Isles, more males than females were buried in these preeminent spots, pointing to an interesting sex bias. In accordance with this observation, the descent of most individuals with kinship links could be traced through the paternal line. In one case it was possible to find two related males buried in two different megaliths two kilometers apart (Primrose Grange and Carrowmore in Ireland), indicating a geographic expansion of these dominant families. Genetic analyses of skeletal remains discovered within the most intricately constructed chamber of the Newgrange passage tomb revealed that they belonged to the incestuous son of a brother and sister (or a parent and child), and therefore a quarter of his genome had no genetic variation.[18] This kind of first-degree offspring is extraordinary, only having been cited in royal families of the past headed by god-kings such as the Egyptian pharaohs seeking to maintain a pure dynastic bloodline. (It is known, for instance, that Akhenaten married his eldest daughter, Meritaten, and much later, Ptolemy II married his sister, Arsinoe II—hence his nickname, "Philadelphus" or "sibling loving.") It has been suggested that this Neolithic elite may have claimed to possess divine powers to ensure the continuity of agricultural cycles by keeping the sun's movements going.

The findings support the notion that these Neolithic communities were socially stratified and that the massive stone structures were used to bury transgenerational patrilineal members of these clans. Perhaps equally interesting is the fact that in one case relatives were separated by up to twelve generations, pointing to an unusual stability through time of both the funerary tradition and the stratified society where they lived.

One of the most illustrative examples of how the analysis of Bronze Age individuals that lived through continental-scale cultural changes can shed light on the process is a study led by researchers at the Max Planck Institute in Jena and published in 2019. Paleogenetic researchers analyzed more than a hundred skeletons from forty-five farmstead-related graveyards in the Lech River valley in southern Germany to explore the social mechanisms underlying the local spread of steppe ancestry across Europe. Additionally, isotope data were generated for these individuals to gather information on their lifetime mobility patterns, which could be correlated with differential composition in genetic ancestry.[19]

The first obvious result was that individuals from the same burial site showed more genetic similarities among them—and in many cases, close family ties along the paternal line—than those from nearby sites, thus indicating that the basic political, economic, and social units were extended

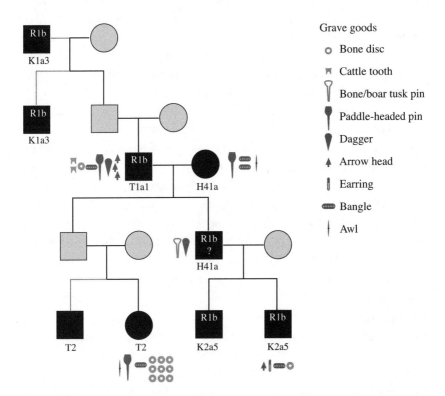

Figure 3.2
Patrilineal social structure in Bronze Age sites from the Lech Valley (Germany), as deduced from genetic analysis. This is one of the longest pedigrees to be reconstructed from these sites, with six generations of the same family identified (*squares*: males; *circles*: females). The males are all descended from the same male ancestor, while females come from outside the original group (hence there is a significant diversity in the mitochondrial DNA lineages—T2, K2a5, and H41a—while all males belong to the same lineage—R1b). Various funerary goods, such as bracelets, arrowheads, and daggers, found within the graves, are depicted. *Source*: A. Mittnik, K. Massey, C. Knipper, F. Wittenborn, R. Friedrich, S. Pfrengle, M. Burri, et al., "Kinship-Based Social Inequality in Bronze Age Europe," *Science* 366 (2019): 731–734.

nuclear families comprised of independent farming households from across the region.

The genetic results also revealed that individuals linked to the early Corded Ware culture inhabitants of the Lech River valley had a higher proportion of steppe ancestry, especially preeminent in their autosomes, than subsequent Bronze Age individuals (prior Neolithic individuals had none at all). Most of the male individuals analyzed carried the predominant Y chromosome steppe haplotype R1b. A plausible explanation for the pattern observed is that incoming Corded Ware migrants with steppe ancestry mixed with local farmers of Neolithic-like ancestry, thus decreasing the former component in subsequent generations.

Isotopic analyses revealed that females tended to be nonlocal (only 50 percent of them had values consistent with the local isotopic range) as compared to males and children from the same cemeteries (where 82 to 84 percent were deemed local). Isotopic data on early and late forming teeth in the same individuals—the first and third permanent molars that emerge at six and eighteen years, respectively—suggested that females moved from their birthplaces during adolescence or later. One of them was found to come from a place at least 350 kilometers away, probably in an area dominated by the Únêtice culture rather than by the Corded Ware one. Most of the males carried the R1b Y chromosome lineage, while the mitochondrial DNA lineage composition was much more diverse. The results indicate that these Bronze Age settlements followed patrilocal residential rules—that is, males stayed in the groups where they were born, while females moved away from them. The fact that most males' descendants shared their ancestry with a single female also suggests that the social structure, besides being based on patrilineal links, was likely monogamous.[20]

Kinship patterns were analyzed between individuals, taking advantage of the fact that first-degree relatives—whether parent-offspring or siblings—share 50 percent of their genomes—a fraction that halves with each generation (25 percent between grandsons and grandparents, etc.). To estimate the degree of kinship, it is necessary to estimate first the genetic variation between unrelated pairs of a population; in small and endogamous populations, two unrelated individuals can have a significant number of matching variable positions across the genome without being closely related. Once this basal population parameter has been estimated, several degrees of kinship can be ascertained, sometimes up to fourth-degree relatives depending on the size of the population and the quality of the data.

The researchers were able to reconstruct six pedigrees in different graveyards, three of which spanned at least four generations. They detected ten parent-offspring relationships, six of them between mother and child. Interestingly, the latter were always male; there were no adult daughters present. Again, this suggests that females were interchanged between households as a way to establish alliances; it is likely that their status was secured once they had children in the new household. It was also possible to correlate grave goods (daggers, axes, chisels, and arrowheads for males, and body ornaments such as neck or leg rings for females) with kinship. This indicates that wealth and social status were inherited and ran with families. The fact that even children who died in infancy were buried with grave goods suggests as well that their status was inherited rather than acquired during their lifetime. A further observation was that members of each clan were buried near each other in the cemeteries, thus clearly delimiting preeminent areas within them. It is likely that the inheritance system of these households was based on male primogeniture—a custom by which the oldest son inherits all the family's properties at the father's death. With time, forged alliances granted families access to larger, regional clans—and eventually kingdoms.

Other contemporaneous individuals were found to be local by isotopic analysis but unrelated to the dominant families; the males among them were buried with few or no grave goods, while females displayed some ornaments. These low-status individuals—who might be considered servants or slaves—showed no obvious differences with the high-status core family in terms of general ancestry composition. The fact that they shared graveyards indicates that they probably lived together in the same households in the Lech River valley. Their labor would have contributed to the wealth of the household with little or no benefit to themselves, at least judging from their social status as deduced via archaeology. Where these "outsiders" came from is a matter of debate; young males from the elite groups might have formed pillaging parties that raided distant communities.

The same pattern of paternal kin and female mobility was found in two Late Copper Age Bell Beaker culture sites (Alburg and Irlbach) from Lower Bavaria in southern Germany, from which four to six generations of two family groups were reconstructed. The Y chromosome showed an extraordinary uniformity; like in the Lech River valley, all males belonged to the R1b lineage. In stark contrast to the paternal line, mitochondrial DNA genomes were highly diverse, with twenty-three different mitochondrial lineages among thirty-four sequenced individuals.[21]

In a different study of ninety-six almost contemporaneous individuals from Switzerland, researchers were able to reconstruct the ancestral composition of the population, detecting the arrival of the steppe component to the region as early as 2800–2400 BCE. Although in the first centuries Neolithic Y chromosome lineages such as G2a and I2a coexisted with the typical steppe lineages, after 2200 BCE all Y chromosomes found are either R1b or R1a. The researchers were also able to reconstruct some family trees spanning three generations at three different sites, namely Oberbipp, Aesch, and Singen. The conclusions drawn from the genetic patterns were similar to those observed in the sites from southern Germany: men tended to be buried with their fathers, brothers, and sons, while only four women were buried together with their families, again suggesting a widespread female mobility. Interestingly, they also found several women without any steppe ancestry when this was already prevalent in the region, with one of them being dated to as late as 2200–2000 BCE—that is, as many as eight hundred years after the arrival of this component in central Europe. Again, the isotopic composition of this woman indicated that she was not of local origin. This suggests that there were either population pockets without steppe ancestry across central Europe, or perhaps women were migrating from distant places where this ancestry had not yet arrived or was scarce; plausible places would be the south of the Italian peninsula or even Greece.[22]

In yet another study, researchers examining twenty-four Bronze Age skeletons, with slightly more recent specimens (between 2100 and 1800 BCE) unearthed at the site of Mokrin, in present-day Serbia, detected nine family ties among fifteen of these individuals. Again, the genetic results showed family links among males with wealth, as deduced by the presence of jewelry such as head ornaments and necklaces. In three cases, the first-degree relationship was between mother and son; in one of these, the son was buried with extravagant grave goods that surpassed those of his biological mother in value, suggesting that social status might be acquired during one's lifetime rather than solely through inheritance. (Nevertheless, some infants, both male and female, displayed signs of high social status that, due to their age, must have been inherited.) An absence of biological daughters at the site—as well as several girls and women with no biological relatives found—again suggests the existence of female mobility between these Bronze Age groups. In accordance with this observation, there were fourteen different mitochondrial DNA lineages in the sample, but only five Y chromosome lineages, with three males belonging to the R1b marker from the steppes.[23]

Finally, the recent genetic analysis of ninety-six individuals from Bronze Age Iberia, including sixty-seven and ten from the famous El Argar sites of La Almoloya and La Bastida, respectively, revealed also a patrilocal and patrilineal society. When pedigrees could be reconstructed, the researchers uncovered several generations of males buried in preeminent, intramural places. In contrast, there was a complete absence of first- or second-degree relationships between adult women.[24] At least in one case, the offspring of the man from a double, princely burial, but with a different, undiscovered woman was identified; whether she was a concubine or a second wife after the first one died is impossible to know (notice, however, that in the second scenario, he chose for some reason to be buried with the first one).[25]

While awaiting more, similar analyses of cemeteries from this crucial period, we can speculate that the evidence for patrilocality and patrilineal inheritance seems to follow a similar social development, at least in central and western Europe. And the general consequence of this pattern was a homogenized distribution of the maternally inherited genetic marker (the mitochondrial DNA lineages) over large geographic areas with a strong influence on the regional structuring of the paternally inherited Y chromosome.

Interestingly, as we are clearly dealing with differential mobility, we can also take advantage of biochemical techniques that provide information on the lifetime movements of individuals. The combination of genetics and archaeological techniques enables us to grasp the microhistories underlying large population changes. In the case of the Amesbury archer, for instance, an isotopic analysis suggested that he grew up in central Europe, probably in Switzerland, Germany, or Austria, and genetic tests confirmed that he had a substantial amount of steppe ancestry—albeit not as much as other contemporaneous continental Europeans.[26] He was certainly not British born but instead a first-generation migrant who had risen to preeminence by the time of his burial near a monumental structure—Stonehenge—originally erected by the predecessors of these Bronze Age newcomers. Young males could be long-distance migrants, but women had higher regional mobility rates once the new social structures were established; in the Alburg and Irlbach cemeteries from southern Germany, for instance, six out of the eight nonlocal individuals—as determined by isotopic analysis—were women, reinforcing the evidence for a higher female mobility in these communities.[27] The genetic finding of a father and daughter—with a neonate—separated by 6.5 kilometers in the English sites of Amesbury Down and Porton Down, respectively, could in fact be an example of this mobility pattern.[28]

In a different archaeological context, the western Mediterranean, a set of archaeological and genetic factors have recently shed light on the existence and organization of privileged minorities among Punic settlements with funerary hypogea chambers. Although the original tombs were looted long ago (ostentatious sepulchral chambers or valuable sarcophagi were generally reused in antiquity), these extravagant structures still stand as indicators of past social status.

The ancient city of Baria in southeast Iberia (now Villaricos in the Almería Province) was founded circa eighth century BCE by Phoenicians from Tyre and remained under Carthaginian control until it was conquered by Romans during the Second Punic War. Baria is well known for a set of monumental funerary hypogea excavated in a rocky outcrop outside the ancient city. Some of these structures, which included a corridor and funerary chamber with some niches aligned along the walls, required a great deal of labor to erect. The site was excavated first by pioneer archaeologist Luis Siret (1860–1934) starting in 1890 and was subsequently followed by more excavations through the first half of the twentieth century, in which 1,842 graves were uncovered—most outside the hypogea. The associated materials, both the skeletal remains and accompanying grave goods, such as ostrich shells, ceramics, and ivory plates, were deposited in the Museo Arqueológico Nacional (Madrid) in 1935 and remain there today. The most prominent tombs were likely associated with the ruling elites of Baria. A set of bones from some of them that have been recently analyzed genetically reveal that the individuals had varying degrees of Iberian or North African ancestry—the latter component likely associated with Punic settlers.[29] The results of one particular hypogeum were quite interesting, as they revealed five individuals, dating from around 400 BCE, who were genetically indistinguishable from modern Aegean Greeks; all were second- or third-degree relatives and clearly inbred (long runs of their chromosomes have no genetic variation, indicating that their parents were related). In addition to nineteen ostrich eggs—luxury items in antiquity—this hypogeum contained two small engraved ivory plaques, one depicting an Ionic capital and the other a Greek mythological theme. The genetic results confirm that this hypogeum was a family pantheon, in use for at least fifty years, and suggest that the Punic city had a small, endogamous Greek community (much further southwest than was previously known). It also represents a paradigmatic example of how genetics can shed new light on archaeological artifacts retrieved from an old excavation that can still be correlated with family connections and ancestral origins across the Mediterranean. It is noteworthy that other samples from the same site, both from common graves and other hypogea, don't show this sign of Aegean ancestry.

Moreover, the fact that Punic and Greek communities coexisted in some Punic cities had political consequences, and underpinned Hannibal's strategy to get the Greek world on his side. (Although he was not always successful in this approach, cities such as Syracuse and Tarentum as well as the kingdom of Macedonia became his allies during his struggle against Rome.)[30]

The same approach of combining genetics and archaeological findings can be applied to more recent, historical periods, especially where migrations were supposed to have taken place, such as the Barbarian movements at the end of the Western Roman Empire—also known as the Migration Period, for a reason. A similar study was conducted in two other cemeteries—one in Italy and another in Panonia (present-day western Hungary), both dated to sixth–seventh century CE.[31] The human remains are dated to the period of the establishment of a Barbarian kingdom—that of the Longobards (literally "Long beards")—in northern Italy. A common archaeological discussion has centered around the ethnic or cultural nature of these Barbarian groups, some of which are cited by Roman chroniclers as simply disappearing from subsequent events, seemingly replaced by other, new groups. Some scholars propose that these groups had a "liquid" identity consisting of opportunistic and ad hoc groupings of diverse peoples, while others claim that they were ethnically—and even genetically—homogeneous groups.[32] Debates about the importance of the demographic movements of this period, which in many ways presaged the formation of modern European states, continue to rage among scholars, with some of them, such as Peter Heather from King's College London in his book *Empires and Barbarians*, arguing that migrations did in fact play a relevant role.[33] (There is also a considerable debate on how to name archaeological complexes from this period without using cultural or ethnic terms such as "Longobard.")[34]

The genetic results from cemetery overlapping with the Longobard period, at a site called Collegno near Turin, demonstrated affinities with central and northern Europeans, consistent with the north-to-south population movements at the end of the Western Roman Empire. Interestingly, the arrangement of both the Italian and Hungarian cemeteries follows kinship connections. In Hungary, a major pedigree spanning three generations and ten individuals was uncovered; many of them had grave goods, but they were also buried in elaborate ledge graves that are otherwise uncommon. Most of the individuals were male and showed nonlocal isotopic signatures, which suggests that this was a migrating unit organized around a high-status, kin-based male group. The women who were found surrounding the male tombs in a half ring of burials showed local isotopic ranges and more diverse genetic composition, indicating that they might have joined the migrating group along the way.

A similarly large family group was found at Collegno as well, again buried with grave goods in elaborate tombs that in this case were dispersed across the cemetery. Individuals with a more southern European ancestry—potentially from a local substratum, an observation also supported by their isotopic signatures—are scattered across the site and display more modest burials. In addition, their protein consumption was lower than that of the members of the large, high-status pedigree. The researchers believe that Collegno might illustrate how incoming Barbarian families exerted a dominant social influence on a resident, late Roman population. It must be emphasized that without having revealed the genealogical links between various extravagant tombs across the site, such a deduction would have been based on pure speculation.

The analysis of a number of skeletons from the Lazio—the region around Rome—showed that during the Migration Period, the individuals underwent a substantial compositional shift in ancestry toward central and northern European populations in a process that continued into the Middle Ages to finally form the modern-day Italian population.[35] These results

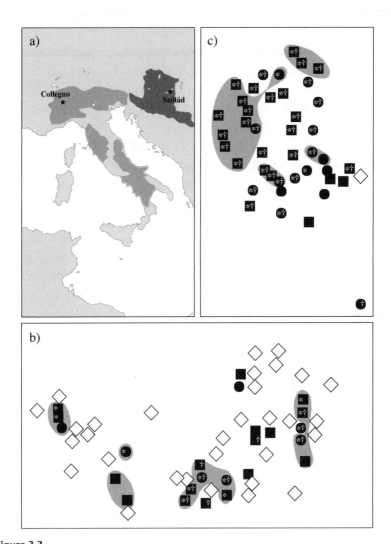

Figure 3.3
Archaeological and genetic characterization of two Late Antiquity sites: Szólád (Hungary) and Collegno (Italy). Map of Europe (a) showing the location (*stars*) of the two cemeteries and regional context (the Roman province of Pannonia in Burgundy and the Longobard Kingdom in beige). Also shown is the spatial distribution of the graves at both sites (b: Szólád; c: Collegno) as well as tombs with grave goods (*) and elaborate construction (†) (*squares*: males; *circles*: females; *diamonds*: undetermined). Shaded areas represent detected kinships. *Source*: C. E. G. Amorim, S. Vai, C. Posth, A. Modi, I. Koncz, S. Hakenbeck, M. C. La Rocca, et al., "Understanding 6th-Century Barbarian Social Organization and Migration through Paleogenomics," *Nature Communications* 9 (2018): 3547.

support the microregional findings from the Longobard period cemeteries in regard to a certain demographic and social replacement after the collapse of the Western Roman Empire on a larger populational scale.

But an examination of the dynamics between kinship and social inequality can be applied to even more recent periods. The complex interactions underlying extended families and population levels can be better understood in geographically isolated places such as islands. Iceland remains the most studied island from a genetic point of view, mainly due to the efforts of a private company called deCODE Genetics that was founded in 1996 by neurologist Kári Stefánsson.

Iceland, a remote island in the north Atlantic, was first colonized around 874 CE, according to the *Landnámabök*, or "settlement book," when the Norse chieftain Inólfr Arnarson arrived in the region of present-day Reykjavik. Over the next 150 years, groups of Viking migrants from Norway along with Celtic women and servants or slaves arrived on the island, establishing themselves on rather isolated farms. By 930 CE, all arable land was already occupied and all the forests were gone. The migratory influx slowed down afterward and almost ceased after the year 1000 CE. This resulted in a population that was small and isolated—yet at the same time big enough to have all the common European diseases and genetic diversity—and it suffered several demographic bottlenecks associated with volcanic eruptions, famines, and epidemics of the plague. Until 1850, the Icelandic population never exceeded fifty thousand. The combination of two factors—an isolated population and a well-known genealogical database—makes Iceland an ideal laboratory for detecting genetic variants associated with common diseases that affect not only modern Icelanders but also the rest of Europe, where such information does not exist or the population is too big to make such an approach practical. Over the years, researchers from deCODE Genetics have generated a whole body of data on the genomics of modern Icelanders and also on how the original population was established. By working with uniparental markers from living Icelanders it could be observed that 62 percent of the mitochondrial DNA was Celtic in origin (meaning that the majority of these maternal markers derived from either the British Isles or Ireland), while 75 percent of the Y chromosomes were of Scandinavian origin.[36] This suggested a settlement primarily established by Viking males and Celtic females.

In 2017, and thanks to paleogenomic techniques, it was possible to retrieve twenty-seven ancient Icelandic genomes, most of them from the heathen period (prior to the year 1000 CE, when Icelanders decided to become Christians by the curious procedure of voting).[37] At the nuclear genome level, these pioneers had a Norwegian-type ancestry (55.4 percent) that was greater than the Celtic one, and more prevalent among men (a recent genetic study of more than four hundred Viking individuals has confirmed the spread of Norwegian ancestry mainly across the North Atlantic islands).[38] Modern Icelanders are not, however, a simple mixture of the two components; their ancestry demonstrates a differentiation from the two source populations at least partially due to genetic drift—promoted by geographic isolation—during the last thousand years. Interestingly, the Norwegian-type ancestry component in Iceland is nowadays 70.4 percent, suggesting an increase that was likely socially mediated. An example of this stems from seven individuals excavated in 1964 from a boat grave (a type of burial in which a ship is used as a container of the dead) at Vatnsdalur in the remote western fjords. The grave goods included a knife, thirty beads, a silver Thor's hammer, a Cufic coin (dated circa 870–930 CE), and various items of jewelry. Three of the four skeletons sequenced showed mostly Scandinavian ancestry. One of these individuals is among the few sequenced early settlers to be genetically similar to modern Icelanders, indicating that he contributed disproportionately to their ancestry.

It seems that the Celtic servants brought to Iceland clearly had fewer opportunities to reproduce. Using isotopic analysis, it was also possible to detect that at least three people—two Scandinavians and one Celt—were first-generation migrants, having spent their childhood outside Iceland. One individual had mixed ancestry, indicating that his parents were from different places. The fact that the Celtic ancestry is still detectable decades after the first settlement also suggests that some kind of social discrimination between the two ancestral groups persisted for a while. After a few centuries, however, the admixing of the two communities was complete, to the point that Iceland has essentially become an extended family with a remarkably uniform population.

We have seen several case studies of past inequality correlating funerary archaeology with genetics that might no longer apply today, where legal regulations (and also the exponential increase of cremations) represent a certain degree of standardization in funeral practices. Nevertheless, an opposite trend could shape the future of the archaeology of death: the trend toward personalized coffins, unconventional funerary memorials, and special grave goods.[39] One way or another, mortuary archaeology will always be an important subfield of this discipline, and one that will need to rely on the hard sciences such as genetics and forensics.

Perhaps one encouraging conclusion is that despite what we have seen on the archaeology of past inequality, societies have been able to evolve and change their social stratifications. One example is Iceland itself; the country has become one of the most egalitarian societies in the world. In 2018, Iceland passed a law that all companies employing more than twenty-five people will have four years to ensure gender-equal payment because, according to the head of the Equality Unit at Iceland's Welfare Ministry, "equality won't come about by itself, from the bottom up alone."[40]

4   How Social Structures Influence Genetics

Caste is a state of mind.

—Bhimrao Ramji Ambedkar, *Writings and Speeches*

The late Roman scholar and stateman Boethius noted in his *The Consolation of Philosophy*—written while awaiting execution—that "the wealth which was thought to make a man independent rather puts him in need of further protection."[1] And this observation is still obviously right; human societies, past and present, have typically sought to establish social mechanisms to maintain the status quo over time as well as justify their inequalities in terms of wealth, power, and prestige. Moreover, Piketty, in his 2019 book *Capital and Ideology*, reaches the conclusion that inequality is primarily dependent on political and ideological factors, and not on intrinsic economic or technological ones.[2] High-status positions are by definition scarce as compared to the number of people striving to climb the social ladder. These positions tend to be permanent and inheritable—or such is the ambition of the upper class. In some societies such privileged positions can be acquired (that is, achieved over the course of one's lifetime through an established mechanism such as a presidential election or successful business career) or assigned (by birthright, as in a monarchy). Individuals in unequal societies are in some way classified and ordered by social position, and with few exceptions such positions remain unaltered throughout their lives. Thus maintenance of the social order is crucial to the persistence of the original inequalities; the resulting genetic evidence that I've examined in this book would not be possible without the social systems that provide support for inequality—sometimes over long periods of time. Even if the recent decades of globalization have instilled in us the idea that all societies are now quite similar, this was not the case in the past and in fact is not even so today.[3]

Thirty years ago, during a trip to Israel, I visited a kibbutz in the Golan Heights. At that time it seemed anachronistic to me, like a tourist attraction. Even so, it *was* a real community, albeit small, based on agriculture and situated in a militarized zone. (Some of its members patrolled the settlement's boundaries with automatic rifles slung over their shoulders.) Kibbutzim are collective communities; the first ones were established at the beginning of the twentieth century. One hundred years later, there are about 250 across the country, with some 143,000 inhabitants in total. The members explained to me that all property, including the houses and even the clothes, belonged to the community and that all workaday issues were discussed at public assemblies. What struck me the most was the fact that all the children were raised together and even slept in the same room (a measure that was conceived as a way to free women from the burden of caring for them; the practice has been abandoned by all the communities). In my view, the social model had somehow failed; most of the children had left this kibbutz as adults to settle in the big cities (though this could coincide with the general trend toward rural exodus). Kibbutzim in present-day Israel have reinvented themselves, and many are now private (including the first one, called Deganya, established in 1910), and are devoted to technological and industrial activities, such as manufacturing cars and military components. In accordance with the rest of the world, the social norms have been redefined, and are now more individualistic and less communal.[4] During my visit, however, I was surprised to discover a clearly different social model, rooted in rational thinking on egalitarianism derived from utopian socialism.

The perception of human societies as unfair realities generating discontent, inequality, and conflict is extremely old. Since classical Greece, philosophers and intellectuals have devoted effort to imagining utopic societies—generally in a single city since it would take a lifetime to plan a whole world—that could steer contemporaneous social systems toward more harmonious models based on principles of equality, freedom, and fairness. These hypothetical worlds, where society is depicted as obeying rules quite unlike those we're familiar with, captivate our imagination, just as the Israeli kibbutz fascinated me.

Utopias are social constructs that explore alternatives to current societies that are purportedly better for their inhabitants. Strangely, most of these proposals share certain traits that may strike most of us as disheartening, not only because of the compulsory norms, but because the premises for equality interfere with freedoms, and such limitations engender new forms of inequality. If this paradoxical situation shows us anything, it is how difficult it is to establish real egalitarian principles, particularly when formulated with almost complete ignorance of fundamental economic, social, biological, or psychological aspects of human nature as well as taking for granted the underlying idea that humans are perfectible.[5] As rational constructions of an idealized world, utopias have rarely materialized—and when they have, they usually end up in a catastrophic state.

The Republic of the Greek philosopher Plato, written about 360 BCE, is considered the first utopia and probably his most famous work. In it, the inconveniences and limitations of the Athenian democracy—to which the unfair trial and death penalty of his master, Socrates, could be attributed—are expressed through dialogues about the relative merits of being just or unjust as well as the philosophical bases on which to build a new society. To our eyes, though, this new society appears in an unfavorable light of rigid organization and discipline. It was to be organized under the following precepts: everyone would live together, there would be no private property, and everyone would have a trade. The ruling class would be formed by philosophers, and the state would be protected by an armed body—the so-called guardians—who were to be raised under specific moral rules that turn them into fine, upstanding citizens. Women, considered by Plato—and the misogynistic classical Greek society in general—as inferior human beings, would be shared among the guardians (we read "the best men must have sex with the best women as frequently as possible, while the opposite is true of the most inferior men and women"), and the children born by them would have no specific fathers. Nevertheless, with the abolition of private families, girls could opt for an education or even perform the function of ruling the society.[6] This one aspect is a rare exception among Western utopias, which invariably place women in a subordinate role, and *The Republic* has even been dubbed—probably overenthusiastically—as "feminist."[7] In fact, the proposal is primarily concerned with the welfare of the whole city rather than that of women in particular.[8]

Another element common to many utopias appears in this first version: strict demographic limitations. Plato's exemplary polis would have an ideal number of 5,040 citizens (among other special mathematical properties, this number can be divided by all natural numbers from 1 to 12, with the sole exception of 11). In any case, this figure is remarkably small for present-day cities and is more appropriate for a village.

The father of modern utopias—and coiner of the word—is the humanist Thomas More (1478–1535). In his book *Of a Republic's Best State and of the New Island Utopia* (known for the sake of simplicity as *Utopia*), we read that everyone will need to learn a profession, including children. Everyone will wear the same loose-fitting clothes, which are supposed to last seven years, with the only variations being to distinguish men from women and single from married people. Work will be limited to just six hours a day—three before lunch and three after—and eight o'clock will be the compulsory bedtime. Sons will obey their fathers, youngsters will obey adults, and women will obey their husbands; the subordinate role of women, their main task being to ensure reproduction, is a common feature of many utopias. Rather surprisingly, wives could participate in battles alongside their men (in stark contrast to Tudor England's conceptions of masculinity).[9] There would be six thousand families (each comprising ten to thirteen adults), distributed across the fifty-four cities of Utopia; as with other utopias, there is a pedantic precision to the demographics.[10]

Yet another influential utopia of the Renaissance was *The City of the Sun*, the work of Dominican friar and philosopher Tommaso Campanella (1568–1639). He wrote it in 1602 while in prison for promoting an insurrection against Spanish rule and his heterodox views, which ran afoul of the Inquisition. For Campanella, the urban organization of the city was of utmost importance. His City of the Sun would be placed atop a high hill dominating a plain and surrounded by seven walls, with a round temple dedicated to the sun in a central square. All inhabitants would work four hours a day—a slight social gain over More's utopia—sleep and eat communally, and share goods, women—a common fantasy of many utopias—and children. Campanella offers a reproductive program that can be classified as eugenic: the production of offspring must be planned for in advance, with special regard to the physical and moral qualities of the parents. Love has nothing to do with the matter. Having children would be a question of social responsibility; people would reproduce for the sake of the community. (According to Campanella, the utopian citizens "deny . . . that it is natural to man to recognize his offspring and to educate them.")[11] Thirty years after writing it—twenty-seven of which were spent in prison—Campanella himself considered whether some of the passages in the book should be modified, but concluded that no amending was needed. This is not surprising as utopias are essentially static affairs and thus lack the flexibility to adapt to unforeseen social changes.

These utopias were not the only ones of the Renaissance, a period of social change that saw the end of feudalism and the emergence of a market economy. In *Christianopolis* (1619), German theologian Johannes Valentinus Andreae (1586–1654) launched the first utopia with a Lutheran flavor, in which the poor, beggars, thieves, and fanatics would be prohibited from entering the community.[12] All types of violence, including the death penalty, would be suppressed, and husbands could only bite their wives occasionally(!). Versatile philosopher and scientist Francis Bacon (1561–1626) left an incomplete utopian novel titled *New Atlantis*, published after his death. In this book, a European ship sails to an island in the Pacific Ocean to find an unknown civilization. Bacon describes its social organization: although the community is—surprisingly—Christian, it is a society based on scientific observation and experimentation; its main goals are comprehension of the natural world and social improvement via technological advances. As in all humanistic utopias, the inhabitants of the island are chaste, honest, and pious, and only procreate as a sort of obligation to the community.[13]

We can perceive an authoritarian flavor emanating from the utopias of More, Campanella, Andreae, and Bacon. In all of them, individual freedom bows to the needs of political organization; happiness in some ways becomes an obligation. Perhaps this is not surprising when we consider that as in Plato's time, slavery was still a legal institution. Modern readers may find it baffling that all their cities' inhabitants are happy as this goes against common sense—yet another paradox found in most utopias.

The nineteenth and twentieth centuries saw the publication of new utopian proposals. Charles Fourier (1772–1837) was a French philosopher and early socialist who challenged capitalism and proposed a new model for society in his book *The Social Destiny of Man* (1808). Fourier advocated for a society structured into cooperatives and self-governed units of fifteen to sixteen hundred individuals that he called "phalanxes." He proposed that one-third of the population should be celibate, and that at least seven of every eight people be dedicated to manufacturing and agriculture. Individuals not contributing to the progress of their phalanx should be expelled from it.[14] While other utopian socialists made similar proposals, Fourier's are remarkable as they were actually implemented in several communities in the United States, such as the Phalanx of Sodus Bay (1844–1846), Wisconsin Phalanx (1844–1850), and Clermont Phalanx (1844–1845).

This, then, was the first time that a utopian ideal was transformed from a theoretical memorandum into a practical matter that affected real people's lives. What all these social experiments have in common is that they ended in failure. The attempts to create utopian communities have been most frequent on the American continent, probably for the same reasons that religious communities escaping from prosecution in Europe were among the first to settle there: the possibility of starting something new without any preexisting social or legal constraints (not in vain was it called the New World). The very foundation of certain states was clearly utopic; in the initial formation of Pennsylvania, for instance, each landlord had to possess the same area of arable land.

Several utopian communities migrated from Europe to settle in the United States; by 1840, there were about eighty of them. Some may still raise a few eyebrows, even by the rather relaxed social standards of today. The community of Oneida, New York (between 1848 and 1881), for example, considered all of its members married among themselves, under a concept they called "complex marriage." New Harmony, a community established in Indiana by social reformer Robert Owen, was based on principles of education, science, technology, and communality. Overcrowding, disputes over credit, a shortage of craftspeople, and the impossibility of becoming self-sufficient eventually led to its demise.

There were other such social experiments in South America. The Jesuit missions established in the most inaccessible regions of Paraguay and northeastern Argentina (in a state still called Misiones) were a kind of utopian experiment based on the creation of self-managed Christian communities; King Charles III of Spain chose to end them in 1767. And the sister of philosopher Friedrich Wilhelm Nietzsche, Elisabeth Förster-Nietzsche, and her husband created a community in 1887 in the jungles of Paraguay called Nueva Germania, formed by fourteen German families. The underlying idea had a kind of Darwinian perspective; under the assumption of superiority of what they called the Aryan "race," the community would rapidly thrive and expand throughout the entire American continent, replacing the original inhabitants. Obviously this did not happen; the land was not suitable for agriculture, disease spread, and the experiment failed after a few years, when Elisabeth abandoned the colony. (Later in life, she joined the Nazi Party.) Nueva Germania still exists, but today it is an impoverished rural community integrated into Paraguayan society through marriage with the locals. (It has been estimated that only 10 percent of their present-day inhabitants harbor German ancestry.) Certain oddities persist; for instance, some families still use German words.[15]

The last century as well saw the practical formulation of new utopias. Yet all the early communities, despite their imagined—or planned—fairness and happiness, ended in failure after several years. German philosopher Karl Marx, who dismissed utopian socialists as unrealistic, argued that a revolutionary class struggle was needed to achieve real change in society. Later on, Communist revolutions leading to the formation of the twentieth-century Soviet Union, North Korea, and China became the dominant socialist movements. Massive deportations, such as those of the Cossacks, Volga Germans, Crimean Tatars, or people from Poland, Ukraine, or the Baltic states are likely to have modified the genetic composition of large European regions. Moreover, it has been estimated that more than ninety-four million people died at the hands of these Communist regimes, effectively turning utopia into dystopia.[16]

The contradictions plaguing these social utopias reveal their deceptive solutions to the problem of inequality, however suitable the diagnosis may have appeared to be. Most emphasize the idea of an egalitarian model, but this is in many ways a subjective concept. We might ask ourselves what kind of equality was sought: of opportunity, human rights, or outcome? Equality under the law, for instance, can tolerate many other inequalities in other aspects of society. Utopias, to some extent, prioritize some over others.

One of the most common aspects of utopian projects is control over reproduction, which as a corollary involves the subjugation in varying degrees of women. Margaret Atwood, in her dystopian novel *The Handmaid's Tale*, published in 1985, depicts a theocratic society of the future, named the Republic of Gilead, that has replaced the United States after its collapse following the Second Civil War. In this society, in which the population is declining because of rising infertility, the few remaining fertile women are forced to have sexual relationships with several men from the dominant elite to bear their children. Their rights are strictly limited, they're denied access to education, and they don't have names. Following the Old Testament, women are assigned different social tasks, distinguished by the color of their clothing; those charged with procreation, for example, wear red.[17] Atwood's book, made into a successful television series in 2017, underscores how utopias end up actually increasing gender inequality. Sally Kitch, a researcher in gender issues from the Institute for Humanities Research at Arizona University, explains that "the only [utopian] communities that managed to achieve gender equity of any kind were celibate societies in which sex and reproduction were just taken out of the equation."[18]

From a philosophical point of view, the uniformity imposed on the utopian models (in jobs, timetables, behavior, and even apparel) is a clear factor of inequality. English philosopher John Stuart Mill had already warned of the dangerous loss of freedom that a society without critics entailed, subject as it was to what he called the tyranny of the majority.[19] Humans are unequal in many respects, including their abilities, interests, and ambitions, and treating them as identical pieces of machinery is unrealistic. The rational search for ideal models to tackle the issue of inequality may actually generate even more unfair societies, likely unable to eliminate the problem.

But what were the social models that utopians intended to leave behind? To comprehend the concerns of the utopian proponents, we have to understand their cultural context first. Classical Greece—and later, Rome—was significantly based on a social class, slaves, that provided cheap labor. Slavery was ubiquitous in ancient civilizations, even if giving slave labor a central role in the economy (in almost all activities, including mining, agriculture, housekeeping, manufacturing, and even commerce) was quite uncommon. It seems that the Greek philosophers themselves—arguably Plato included—found slavery an acceptable institution.[20]

The Renaissance humanists, such as More, Bacon, and Campanella, lived after the end of the Middle Ages and the emergence of urban societies, but before the Industrial Revolution and the subsequent increase in inequality associated with it. In previous times, social differences were accepted as God's design, and later on, as "natural" (Victorian evolutionists even saw these differences within and between populations as consequences of natural selection). Interestingly, using databases of more than 720,000 manuscripts and printed documents between circa 1100 and 1830 CE, and searching for such Latin keywords as *aequalitas/inaequalitas* (equality/ inequality), researchers have been able to reconstruct long-term economic developments and find out how inequality grew from the Middle Ages until the eve of the Industrial Revolution.[21] Another study of wealth distribution in Europe between 1300 and 1800 also demonstrated the trend toward greater inequality since the Middle Ages, with a long-lasting decline associated with the pandemic disruption of the Black Death around the mid-fourteenth century. During this period, the portion of overall wealth owned by the richest 10 percent declined to around 15 to 20 percent; in fact, the richest 10 percent only recovered its previous share around the seventeenth century. In addition, the study showed that society was becoming more polarized between rich and poor through the seventeenth and eighteenth centuries.[22] The social utopians of the nineteenth century were of course immersed in the problems of inequality associated with the Industrial Revolution, which Piketty studied extensively by surveying numerous documents and data. It remains to be seen what future utopians will propose to help address the rampant inequality that we're currently experiencing.

We have seen a continuous rise in economic inequality in the last few thousand years, along with various social models that were to some degree challenged by philosophers and social reformers. But what most interests us is to what extent social models that reflect inequality have consequences for the genetic structure of human populations.

Bronze Age social models had a primary importance in human social history; the authors of the genetic analysis from the Lech River valley cemeteries suggest that those households are similar to the much later *oikos* in classic Greece and the Roman *familia*, both models comprising the kin-related family and their slaves.[23] Piketty argues that post-Roman, feudal societies, which he calls "Ternary Societies," were organized into clerical, military—or nobility—and working classes. This basic social structure can be found in almost any ancient society, from Europe to China, Japan, and also India.[24] Despite the long-term persistence of these ancient social organizations, however, they can no longer be found as they were formulated thousands of years ago. Nevertheless, other social constructs with remarkable levels of inequality currently exist, and one of them—the caste system in India—seems to derive at least originally from the same steppe nomads who shook Eurasia long time ago. And once again we can study their genetic impact.

Probably the paradigmatic example of the world's social hierarchy is the complex caste system of India (although similar structures exist in other societies). In India, there are four main castes (or *varna*) and numerous divisions within them (*jatis*).[25] Jatis are usually identified by their jobs; this would explain why there are about three thousand of them, with more than twenty-five hundred additional subdivisions.

Among the highest caste are the Brahmins, who are responsible for teaching and maintaining sacred Hinduist traditions; next in importance is a warrior and political elite (Kshatriya), followed by commoners who include merchants or moneylenders (Vaishya), and artisans and farmers (Sudra); the two latter groups are somewhat vaguely defined, thereby allowing for a certain mobility between them. Beneath them all is yet another group, called Chandala or Dalit, which includes people categorized as "untouchable," considered so inferior that in many ways they stand outside the system. In the past, they had to work as manual laborers, cleaning streets and latrines, sometimes with their bare hands. These untouchables were banned from drinking the same water as others, and the upper castes even avoided stepping on their shadows. According to a 2007 report by Human Rights Watch, the Dalits still, in this century, have less access to health care, drinking water, and education than the average Indian, as is also reflected in lower life expectancy. As with other discriminatory communities, Dalits are overrepresented in prisons (comprising 25 percent of the Indian population, they make up more than 33 percent of inmates). Moreover, only 1 percent of the highest government jobs are held by Dalits.[26]

One notorious trait of the castes—and to a large extent, the jatis as well—is their genetic impermeability; once you are born into a particular caste—or even into a jati—you are trapped within it. You can only marry someone of the same social level, risking descent to an inferior status if you marry a member of a lower caste. (The rules are somehow more permissive for women than for men.) This creates a remarkable endogamy and generates genetic differences between jatis—who may be living side by side in the same neighborhood. Such differences are larger than those between two random populations at opposite corners of Europe. This endogamy is not without health implications; it seems clear that many medical conditions of a genetic basis in India are related to mutations that are structured between jatis and have risen to high frequencies because of endogamy.

Besides castes, India is comprised of tribes, an official name given to all groups—most of which practice neither Hinduism nor Islam—that are not integrated into the caste system and likely represent ancient population substrata in the region, prior to the social changes that created the caste system. They represent about 8.2 percent of the Indian population (though a minority in terms of numbers, they speak more than seven hundred dialects from different language families).

The caste system is clearly old; it is described in detail in the Manusmriti, a book written in Sanskrit that dates from the second century BCE or maybe a bit later.[27] Some scholars think that thousands of years ago, there were numerous tribal groups in India that were to a large extent endogamous. Imposed by a dominant elite that placed itself in the highest echelon, the caste system encompassed many tribes, with jatis relegated to the lower castes. With time, some jatis were able to climb to higher levels, but the rules for endogamy were strictly maintained. Yet there is some evidence that some centuries ago, the castes were not more rigid than social classes in medieval Europe, and in fact experienced complex social transformations associated with regional military conflicts; if anything, they were used to justify changes in social dominance and thus were in a state of constant but gradual evolution.[28]

This process of renovation also would explain why the relationship between the varna and jatis is complex and variable. In fact, Nicholas B. Dirks, in his book *Castes of Mind*—first published in 1992—argues that the British colonial empire promoted the social importance of the castes and even implemented the system in regions of the Indian subcontinent where it was not especially significant, as a way of "systematizing" India's enormous diversity in terms of social identity, community, and organization.[29] Hence the caste system, now a key symbol of India that encapsulates its essence and uniqueness, was in fact a British strategy to ensure its political dominance—with Britain being a kind of new upper caste atop the original ones. Moreover, the possibility of diluting caste control was one of the factors that underlay the rapid expansion of Islam during the Mogul Empire in the sixteenth and seventeenth centuries. Under Muslim rule, a more relaxed social stratification, partially based on the converts' previous castes, was established. Also, a mass conversion of thousands of Dalits to Buddhism in 2018—previously some became Christians—has been interpreted as an act of rebellion against the system, much to the dismay of the Hindu nationalist parties.[30]

The arrival of genetic studies brought the promise not only of elucidating the original genetic structure of the castes but determining their age too. Several studies prior to the genomic era showed that only a small fraction of the Indian mitochondrial DNA lineages (around 10 percent) were of western Eurasian origin; the rest of the mitochondrial gene pool, which is highly diverse in terms of lineages, is typically found in South Asia. In the paternal counterpart, researchers found the Y chromosome to be much less diverse; more than half belonged to R1a, a lineage also abundantly found in central and northeastern Europe. This western-Eurasian component constitutes the most common Y chromosome lineage in northern regions of the subcontinent—notably in the highest castes (for instance, up to 60, 72, and 68 percent in Bihar, West Bengal, and Uttar Pradesh Brahmins, respectively)—while it is present in only 5 to 30 percent of the paternal Y chromosome of tribal groups (with an estimated overall frequency of around 17.5 percent).[31] Most of the remaining high-frequency lineages—O2a, O3, and H, prevalent in tribes as well as in the south—are autochthonous to South Asia.

Subsequent genomic studies on modern Indian individuals have demonstrated that top castes do have a significantly higher fraction of western Eurasian-related ancestry than the lower ones—reaching up to 80 percent in Brahmins—and that this ancestry follows a geographic cline, decreasing as we progress from the northwest to the south of the subcontinent.[32] The isolated tribes of the Andaman Islands, in the Bay of Bengal, do not show this component. Furthermore, western-Eurasian-related ancestry was 7 percent lower on the X chromosomes, again indicating male-biased admixture at some point in the past.[33]

This differential ancestry is correlated with some phenotypic traits; for instance, the Brahmins have paler skin than the lower castes. It also correlates with language; speakers of Dravidian, more widely spoken in the southern half of India, have less European-like ancestry, as opposed to the speakers of Indo-European languages, such as Bengali, Gujarati, and Punjabi. A pertinent question remains: When was this system implemented?

A recent major paleogenomic study of more than five hundred ancient individuals from Central Asia and India's neighboring regions, led by Reich from Harvard University, showed that Indian ancestry is composed

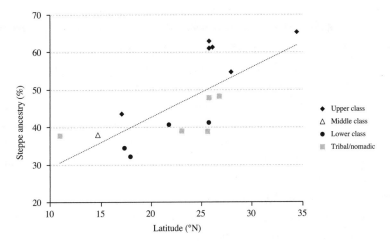

Figure 4.1
Steppe ancestry component (named "Ancestral North Indian" in some genetic literature) in different Indian groups arranged by latitude and caste. Despite the complexity of Indian populations, there is a significantly higher presence of this ancestry in a) upper castes and b) groups located in the north of India. *Source*: Data taken from P. Moorjani, K. Thangaraj, N. Patterson, M. Lipson, P.-R. Loh, P. Govindaraj, B. Berger, et al., "Genetic Evidence for Recent Population Mixture in India," *American Journal of Human Genetics* 93 (2013): 422–438.

of three overlapping layers: an extremely old one, associated with South Asian hunter-gatherers (named "Ancient Ancestral South Asians" by geneticists); a second of Iranian origin, which could be associated with the spread of farming; and a later one, again coming from the steppes and arriving around four thousand to thirty-two hundred years ago in the Indus Valley and beyond.[34] This later component seems to be associated with a movement of people that probably also brought Sanskrit—an Indo-European language—to India and established a precursor to the caste system as a way of consolidating its social position over a vast geographic area with an already large population. Among the 140 modern Indian groups analyzed, the Brahmins showed a strikingly rich component of the steppe ancestry. It is likely that around the second millennium BCE, the migration of groups harboring a large fraction of steppe nomad ancestry played a central role in spreading early Vedic culture.

The same research group published the first retrieval of genomic data from an individual belonging to the Indus Valley civilization—a woman—whose skeletal remains dated back to between 2800 and 2300 BCE. This was a remarkable technical achievement since the climate in this region is very hot and therefore clearly unfavorable to DNA preservation.[35]

The Indus Valley—or Harappan—civilization is, alongside those of Egypt and Mesopotamia, one of the three largest and earliest human civilizations as well as the oldest urban settlement on the Indian subcontinent. It was located on the plains of the Indus River valley—in what is today Pakistan and western India—and flourished at its peak between five thousand and thirty-seven hundred years ago. Some of its cities, notably Harappa and Mohenjo-Daro, thrived with tens of thousands of inhabitants; these cities were built of brick and had technologically advanced urban plans, including granaries, warehouses, docks, and drainage systems under the streets. This civilization's decline and abandonment remains a mystery; some have suggested that it was affected by climate change and episodes of drought—exacerbated by the changed course of the river, which today runs about twenty kilometers away from many Indus Valley settlements. Other historians have related the end of this civilization to the arrival of foreign descendants of the steppe nomads, who have been called "Aryans."

Going back to the genetic study published in 2019, the person from the Indus Valley civilization lacked, as expected, the steppe nomad ancestry that was later prevalent on the subcontinent. It could also be discerned that the Iranian ancestry exhibited by this individual was likely present in previous hunter-gatherer groups from the region that seemingly had started farming. The fact that this and eleven other sequenced individuals from the periphery of the Indus Valley in present-day Iran and Turkmenistan show no steppe ancestry suggests that the arrival of migrants carrying this genetic component could have played a role in subsequent social and political events. In fact, it would have occurred toward the end of this civilization, triggering substantial genetic and social changes across the subcontinent. This seems to correlate as well with the oldest text in India, the Rig Veda, likely composed between 1500 and 1200 BCE, and written in ancient Sanskrit, in which the mythical origin of the castes—formed from the different body parts of Purush, the primal man—is explained. Although neither signs of destruction at the Indus Valley sites nor potential material links with the central steppe Bronze Age cultures have been found, it is plausible that the newcomers had a clear military advantage in the form of horses, as happened in Europe more than a thousand years earlier. Whatever the scenario,

Figure 4.2
Excavated ruins of the large city of Mohenjo-Daro (present-day Pakistan) from the Indus Valley civilization. In the foreground, there is what seems to be a public bath, and in the background, a large granary mound. The urban plan is still clearly visible. Image from Wikimedia Commons, Saqib Qayyum, Creative Commons—Attribution ShareAlike 3.0.

"the lack of material culture connections does not provide evidence against the spread of genes," as Reich explains in an interview.[36]

The idea that western Eurasian peoples could have shaped modern Indian diversity is, quite understandably, viewed negatively in India, where nationalism is a growing ideology. (In fact, the genetic results point to peoples from western Asia, not from Europe, as drivers of this social change.) Some years ago, local scientists supported the view that the existence of an R1a Y chromosome among them was not attributable to a foreign gene flow but instead that this lineage had emerged on the subcontinent and spread from there.[37] But the phylogenetic reconstruction of this haplogroup did not support this view. (So far, however, R1a is absent from the Bronze Age samples from the periphery of India, thereby raising an interesting point about a potential decoupling between the spread of steppe ancestry and the arrival of its characteristic Y chromosome lineage.)[38] The ideological background against which these ideas clash, even today, can be easily related to India's colonial period.[39] For the British, evidence of a foreign migration of western Eurasian origin could be used as a precedent for their political dominance. For the Nazis, the idea of Aryan migration as an engine for the spread of civilization, so popular in the period between the two world wars, served to reinforce their racial conception of superiority and justify their right of conquest.[40]

Genomic analyses have recently focused on the study of genetic diversity within jatis. A surprising result is that these social constructs acted as reproductively closed entities, and to such a degree that their endogamy was genetically remarkable, especially considering that they lived together with people from other jatis. For example, an intermediate jati termed Vysya, which currently comprises about five million people, appears to derive from a small founding group from around three thousand years ago.[41] All its members descend from these few ancestors, to the point that few people within Vysya carry external genetic influences. Genetic differences between adjacent jatis turned out to be about three times greater than differences found between geographically isolated European populations. Thus even if different jatis live in the same building, or their members meet in the street or use the same public transport, they only marry members of the same jati. Moreover, the magnitude of the observed genetic differences means that it is likely that the underlying social divisions have persisted for more than just few hundred years.

The caste system was officially abolished in India with independence from the British in 1947. It nevertheless persists in everyday social relations, partly as a mechanism of social identification; it has even been successfully exported to the Indian community living in Great Britain.[42] Its persistence is partly rooted in attitudes of respect toward the elders of the family, as marriages of family members traditionally need their approval. What is remarkable is the resilience of a social structure that has technically been abolished and its long-lasting genomic effects. Its ability to last into the future is currently a mystery too. Some recent studies documented exogamy (outside marriage) rates to be around 5 percent in rural India and as high as 12 percent in urban areas at the end of the twentieth century. Following basic population genetics estimates, this means that it could take hundreds of years for the caste system to fade away significantly. Thus even in times of globalization, the Indian castes, considered an elaborate form of social stratification and control, still show genetic strength.

India is not the only society organized in castes or closed social groups. There are similar structures of social isolation in African countries such as Mali, Mauritania, Senegal, Gambia, Ivory Coast, Niger, Burkina Faso, Cameroon, Liberia, Sudan, Sierra Leone, Algeria, Nigeria, Chad, Somalia, Ethiopia, Eritrea, and Djibouti. Most of them share such traits as tribal origin along with the establishment of the system between the eleventh and fifteenth centuries, likely alongside the phenomenon of slavery. Many of them are, as in India, associated with specific jobs and restrictions on marriages outside one's own social group. Some of these systems are so complex that it takes time for external observers to grasp how they work. Notwithstanding

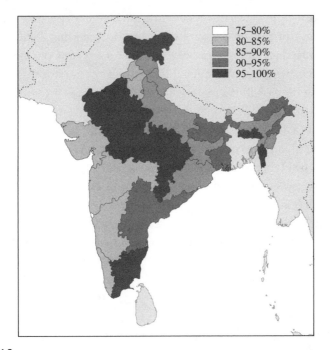

Figure 4.3
Percentage of same-caste marriages in different Indian states at the start of the twenty-first century. The persistence of the caste system is regionally structured, with a partial reversal in the regions with a higher incidence of tribal groups that are outside the system. Notice that in vast regions, intercaste marriages are less than 10 percent of the total marriages. *Source*: National Family Health Survey, NFHS-III, 2005–2006, government of India.

the fact that they are not explicitly written or legally sanctioned, these classifications have a strange consistency for the inhabitants of these African countries. If someone owns a small business, for example, they'll prefer to have members of their own tribe as employees. Future genetic studies on these African communities could reveal how old these structures are and the degree of effective endogamy associated with them. What seems clear is that tribes still act as a barrier to gene flow, even within the same country.

The pre-Columbian Americas were also the stage for specific, autochthonous social structures that can be studied via ancient genetics. Spanish colonial chronicles explain that central Andean populations, notably Quechuas and Aymaras, corresponding to present-day Peru, Ecuador, and Bolivia, practiced a unique social model called *ayllu* (literally, "family clan" or "enlarged community"). Within-group (or ayllu) unions were preferred over between-group ones; this system of close-kin marriages (or endogamy) created a lack of genetic diversity within ayllu somewhat similar to that observed in Indian castes. One way to look for signs of this norm is by searching for stretches of the individuals' genome where both chromosomal copies—one from each parent—are identical (the longer the chromosomal section, the more recent a common ancestor of the parents would be). The presence of these sections is therefore indicative of recent population history and social marriage practices; a large population where mating is closer to random will be devoid of such genetic signs in its genomes. By analyzing forty-six ancient Andean individuals, researchers could detect that after 1000 CE, and prior to the arrival of the Spaniards, the rate of family-related unions underwent a substantial increase, from 9 to 46 percent.[43] This trend coincides with the decline of some widespread archaeological horizons in the Central Andes, which led to a transition to fragmented sociopolitical unities primarily defined by shared ancestry. This shift in turn intensified episodes of intergroup violence prior to the arrival of the Incas, who maintained the ayllu system as a fundamental social element in their empire.

It is interesting to consider that social systems such as the Indian castes and the ayllu may have similar genetic consequences as those observed in small, isolated populations (due either to geography, as on small islands, or religious reasons, such as the Amish) along with the resulting loss of genetic diversity due to endogamy. Thus although populations in caste India and pre-Columbian South America were demographically large, the genetic diversity resulting from their social structure would be smaller than

expected from their numbers alone. As Reich explains, the India of 1.3 billion people would be, from a genetic point of view, the equivalent of a large
number of small populations. The same effect has been recently described
in British Pakistanis, in whom genetic studies have demonstrated extensive
endogamy patterns for the last seventy generations or so (that is, as far back
as fifteen hundred to two thousand years ago, coinciding with the likely
origin of Indian castes), with some major bottlenecks in the last five hundred years, perhaps coinciding with Islamization.[44] In addition, this genetic
pattern is associated with unusually high frequencies of some recessive
mutations that cause serious health conditions due to centuries of perva-
sive genetic isolation in these social strata. With around seventy million
people being affected by rare diseases (defined as those affecting a minimal
fraction of the total population), this represents a significant burden on
India's public health system.[45]

People of certain castes may have very low status, as we've seen, but
there is yet a further step down in social marginalization and extreme
inequality: slavery. In past centuries, Africa was the focus of one of the
most shameful social institutions to be implemented in historical times:
the extensive use of slave labor by European colonizers in the Americas.[46]
It has been suggested that agricultural slavery—as well as serfdom in six-
teenth- and seventeenth-century Russia—was originally triggered by the
opening of abundant free land.[47] Whatever the explanation, a staggering
figure of between ten and twenty million Africans were forcibly transported
across the Atlantic, first to Spanish colonies and later on, notoriously, to
the southern states of the United States, to be used as a cheap labor force,
initially in the production of tobacco and later on cotton plantations. In
the United States, slavery was institutionally planned at levels of detail that
nowadays are shocking; for instance, in Virginia, the Carolinas, and Louisiana, laws enforced between 1820 and 1840 established harsh penalties for
those who taught slaves to read and write.[48]

The existence of slave classes in the past is also central to the purpose of
this book since slaves were almost always outsiders to the population where
they were taken by force and therefore had a different ancestry—often mani-
fested in physical differences, notably pigmentation—that can be examined
via genetics. The physical distinctiveness of African slaves could influence
differential aspects with slavery in antiquity too. For instance, in ancient
Rome, owners and slaves usually had a similar European background, and

this could explain why manumission—the act of freeing slaves by an owner—was quite frequent and socially accepted in that society. In fact, it could also be that the Victorian ideas on African race inferiority emerged parallel to the hardening of Black slavery as a justification for this shameful institution. One can speculate that in the distant past, people's appearance was maybe not that important (we can think, for example, on how the conspicuous pigmentation differences between Mesolithic foragers in early farmers in Europe did not prevent their admixing).

Ancient genetics has begun to shed light on the forgotten ancestry of some of these trans-Atlantic African slaves. A pioneer study retrieved genome-wide data from three slave skeletons—two men and a woman aged between twenty-five and forty years at the time of death—found during construction work in Philipsburg, the capital of the Caribbean island of Saint Martin. (Remarkably, only one of the slave vessels that arrived in Saint Martin between 1650 and 1700 is listed in the Transatlantic Slave Trade Database.)[49] The human remains, dated to between 1660 and 1688 CE, showed sharpened front teeth—a tradition that is still found in some African tribes today. Despite the technical challenges of retrieving DNA from tropical environments such as the Caribbean, the genetic analysis could trace their origins to northern Cameroon in one case, and present-day populations from Nigeria and Ghana in the other.[50]

The same approach was applied to contemporaneous human remains found in present-day Mexico (then known as the Viceroyalty of New Spain). Between 1600 and 1640 CE, seventy thousand African slaves arrived there in response to an increased demand for labor that coincided with a Native American demographic collapse associated with the conquest and spread of imported diseases, including smallpox, measles, and typhoid fever. A recent excavation of a mass grave at the San José de los Naturales Royal Hospital in Mexico City unearthed three male skeletons dating from the fifteenth–seventeenth century CE that displayed artificially modified teeth—a common trait in some Central African tribes.[51] Subsequent genetic results confirmed that they were 100 percent African, with strong affinities to present-day sub-Saharan groups from, or from the vicinity of, Gabon, Cameroon, the Republic of Congo, the Democratic Republic of Congo, and Equatorial Guinea. It was not possible to detect the exact location on the continent, perhaps because of the heterogeneous nature of the original slave trade and because the African genomic information is still quite patchy. Nevertheless, the dating of the skeletons and lack of additional Native American ancestry as well as the results of isotopic analyses applied to the samples suggest that these were indeed first-generation slaves in New Spain. The osteological analysis of these skeletons provides a grim image of these peoples' lives, pointing to hardship and violence. One individual suffered osteological conditions associated with intense labor and physical activity, including gunshot wounds as well as a healed fracture of the right fibula and tibia that deformed his knee. And not only did their skeletons speak of harsh lives, but metagenomic analysis of the DNA teased from their remains yielded signs of the different diseases that they suffered from and maybe even succumbed to eventually. For example, one contained DNA sequences derived from the bacterial species that causes yaws—a contagious skin condition associated with poor hygiene—while another was infected with the hepatitis B virus.

Not only were the living conditions of these African slaves appalling, but having a darker pigmentation than European colonizers meant that the level of visibility and social segregation of New World slavery was extraordinary, even in the context of other, past slave civilizations. This inequality, primarily based on pigmentation, must in turn be reflected in the genetics of modern North, Central, and South Americans, as we will see in chapter 5.

We might wish to think that Western societies are free of castes (understood as rather hermetic social classes), but these societies are certainly socially structured, and economic differences are to a large extent inheritable, as we saw in chapter 1. And even if we conclude that these structures are not as rigid and endogamous as those seen in India, it is also obvious that people do not get married—nor have children—at random within the population. To study such social stratification, we need more subtle genetic tools and much more data, but the effects are likely still operating, even in Western societies.

Several recent studies have attempted to investigate whether genetic differences can give rise to socioeconomic inequalities; as wealth also depends on some environmental factors, such as educational attainment, the conclusions of these studies are sometimes controversial. In a survey of more than 112,000 participants from the United Kingdom, Biobank (a long-term health resource for biomedical research that holds genetic as well as environmental information) researchers compared the genetic composition of a low-income population sample with the wealthiest people in the data set.

If genetic variants turn out to be significantly different between the groups compared, it could be supposed that they correspond to the genetic differences underlying those traits—albeit only if the rest of the traits indicate that the two groups are not structured in other ways, such as by sex, age, ethnic group, or geographic location. The researchers found that genetic variation could account for 21 percent of the variation in social deprivation and 11 percent of the variation in household income.[52] The genes involved in these differences were the same as those involved in mental disorders such as schizophrenia, depression, neurosis, or biological issues like synaptic plasticity. Of course, it is difficult to know whether these results indicate that some genes influence our final social status, or if they represent an indirect influence through personality traits or cognitive abilities that contribute to attaining such status.

In another study with 6,815 subjects from Scotland, the researchers found that genetic variants accounted for 18 percent of the variation in socioeconomic status, 21 percent of the variation in education, and 29 percent of the variation in cognitive abilities, again suggesting that social stratification has an underlying genetic basis.[53] Nevertheless, the message of these studies can be read in just the opposite way: that is, there is a substantial, but not exclusive, environmental contribution to social differences in Western countries. Obviously as a society, we can and must act on these environmental factors. But whatever the reading of the results, it seems obvious that some incipient genetic differences are operating at the social level, and this trend could increase in the future if these social differences grow.

Yet the impact of these differences may to some extent be attenuated due to a tendency for educational level—in turn correlated with income—to be associated with reduced fertility, as reported in another study undertaken with more than 109,000 Icelanders. The researchers found that educational attainment is correlated with delayed reproduction and fewer offspring than in the average population—a trend that is especially accentuated for women.[54] (Incidentally, it is worth noting that if this negative correlation is true, the underlying genetic basis would be declining in the general population because the genetic variants in the causative genes are purged from the population.)

In conclusion, past social structures operating for at least three thousand years in India or those shaping the American populations in the last five hundred years yielded a remarkable genetic signature that is still discernible. But such a link between biology and social structures is not unique to these continental regions. Other subtle cultural and social mechanisms could be operating nowadays, shaping the genomic diversity of human populations, even in Western societies. The persistence over time of some of the structures described here—despite present-day egalitarian values—are the best example of the complex interconnections between society and biology along with their long-term genetic consequences.

5 Gender and Genetics

For most of history, Anonymous was a woman.
—Virginia Woolf, *A Room of One's Own* (adapted)

On September 22, 1663, a contingent of thirty-six young French women arrived at the port of Quebec, then still a village of eight hundred people. These women, known as "Filles du Roy" (literally, King's Daughters), were part of a program sponsored by King Louis XIV as a way to correct the increasing gender bias observed in the French colonies in North America; put simply, there were not enough women to ensure the demographic growth—and thus future—of the precarious colony.[1] Between 1663 and 1673, more than eight hundred young women—mostly from a Parisian orphanage, but also some from the lesser bourgeoisie—made the hazardous Atlantic trip to settle in what was in every sense a remote outpost of the French Empire. Despite the inevitable gossip, they were not prostitutes; in fact, the king provided them with a dowry, and they had the choice to marry or not. Demographically speaking, the project was a success; the population of New France (the French colonial territories including Acadia, most of present-day Canada, and Louisiana—incidentally, named after the same king, Louis XIV) rose from thirty-two hundred in 1663 to sixty-seven hundred in 1672. Currently, most of the seven million French Canadians can still trace their ancestors to those pioneers. (In fact, the fertility rate of those settlers remains among the highest in recorded modern history.) Such famous Canadians and Americans as Angelina Jolie, Hillary Clinton, and Céline Dion can trace their ancestors back to some King's Daughters.

What is interesting about this story is that people in the French court were able to identify a problem—the dearth of women—that was likely prevalent in antiquity as expanding populations settled into new territories. As we have seen in previous chapters, having ancient genomes makes it possible to trace back the arrival of populations with distinct genomic backgrounds and reconstruct past social structures. We also have evidence that many of these migrations were in fact strongly male biased. This means that for these migrations to have a discernible genetic impact on subsequent human populations, interactions with local women had to take place, involving different social and cultural strategies. It must be emphasized that this particular aspect of past migrations could only be approached via historical records or simply from speculation—until now.

Besides this differential male-to-female contribution to current mixed human populations, there are additional factors that generate gender bias. It takes place when one of the sexes has contributed at disproportionate levels to a population (notably, gender balance is nearly 50 percent in natural populations). One of these factors is patrilocality, a mating strategy in which males form patrilineal clans and are less mobile than females; genetic studies have revealed this in some Bronze Age European populations (chapter 3). Another such factor—the two are not mutually exclusive, of course—is polygyny, a social practice in which a man can mate with multiple women. Some researchers speculate that humans have had a long evolutionary tendency toward monogamy—a trend that would explain the relatively mild (around 15 percent) sexual dimorphism in our body size and general robusticity as compared to other anthropoid primates.[2] To be sure, polygynous marriages are still legal in about eighty-two countries—mostly Muslim—across the world. But this matrimonial institution is not just a question of religion; it is in part modeled by economic factors. For instance, in anthropological studies on modern pastoralist groups it has been observed that wealthy men—those who have larger herds—tend to have multiple women and consequently more offspring, in opposition to what is seen in agriculturalist groups.[3]

But we need not think in terms of practices that are illegal in Western countries; notice that what is called "serial monogamy," where men have two or more long relationships, one after another, would likely have a similar genetic impact as polygamy, if men monopolize the fertile period of their partners. (For instance, the famous actor Anthony Quinn had twelve children from four different wives and partners, the first one in 1938 and the last one almost sixty years later, in 1996.) The opposite strategy, where a woman can have two or more partners, is called polyandry. Even though it has been documented in places with scarce land or resources, possibly as a way of limiting the population, this is much rarer in human societies; the best-known example is the Mosuo, a population of about forty thousand people in the northwest Yunnan Province in China.[4]

Mixed populations that had gender biases show contrasting estimates of ancestry when genetic markers with uniparental inheritance patterns (those that are inherited only from either the mother or father) are analyzed and compared. This trend can be observed in mitochondrial DNA, which is a small circular genome transmitted along the maternal line, or via the Y chromosome, its masculine counterpart. The fact that mitochondrial DNA lineages are much more extended across vast continental areas while Y chromosome lineages tend to remain more geographically structured has been interpreted as support for a long history of patrilocality among human groups.

Of particular interest is that gender biases can also be estimated at the genomic level because of the differential contribution by men and women of different origins, due to the simple fact that women possess two-thirds of the entire supply of X chromosomes at a given moment (women are XX while males are XY). If we assume equal male and female populations, the comparison of a given ancestry in the X chromosomes versus the autosomes in the same individuals can give us information about a gender-biased migration carrying that specific ancestry; if, for instance, there were less of this ancestry in the X chromosome it would mean that the migration was skewed toward males. Of course this is not that surprising because even nowadays, many migrants trying to reach Western countries are young males. What is new is that we can now test this potential bias by studying the genomes of people buried in archaeological contexts from contact periods.

One technical limitation is that sexual chromosomes are rather complicated from a structural point of view (especially the male counterpart), with regions almost identical between both X and Y, and repetitive sequences and a few functional genes among them too. In fact, the completion of the first gapless assembly of a human X chromosome, made with the most advanced sequencing technologies, was not published until as recently as

July 2020.[5] This means that only high-quality genomic data can help to unravel genomic gender biases, although obviously it can be complemented with information in the markers inherited via one parent or another. If, for example, women have contributed more to an admixed population than men, the estimates of their ancestry would decrease from the mitochondrial DNA lineages to the X chromosome, from there to the autosomes, and finally to the Y chromosome.

As we've seen, the Neolithic dispersal did not seem especially biased, meaning that both men and women moved across the continent as farming progressed, mixing with the previous foragers. But much later, in individuals from the Corded Ware cultural horizon that emerged after the arrival of the steppe nomads—the Yamnaya—some researchers described a potential gender bias, with the steppe ancestry underrepresented in the X chromosome as compared to their autosomes.[6] The results of this study were misleadingly reported in some newspapers as follows: "In the past, a sole man had children from each of 17 women." If this were literally true, not only would there be a decrease in steppe ancestry in the X chromosome but the diversity in the autosomal genome would also have been drastically reduced, which was not observed in contemporary individuals. Subsequent studies failed to replicate these results, at least using the same methodology and samples.[7] At least for individuals buried in a Corded Ware context and excavated in Estonia, however, an analysis of their X chromosomes revealed less steppe ancestry than their autosomes did.[8] This suggests that the late Neolithic newcomers into the Baltic region previously underwent a sex-biased admixture formed by males with steppe ancestry and females with the prior European farmer ancestry.

Whatever the scenario of the first Yamanya arrival into eastern Europe might have been with relation to sex bias (right now it is likely that we'll need to wait for more data), the subsequent spread of steppe ancestry across Europe and South Asia during the Bronze Age period provides evidence that this process was strongly male biased.

Obviously these observed differences between the Neolithic and Bronze Age indicate remarkably different types of interactions between locals and incomers during the two periods. These differences can be partially attributed to different demographic scenarios; as we have seen, the Late Neolithic populations, despite their internal crisis, were already demographically large, especially if we compare them to the previous foragers. From this alone it is clear that the Bronze Age interactions must have involved more complex social situations.

A mating mechanism that operates when a migrating group is mainly formed by males involves incorporating local women into the incoming group. The genetic analysis of Bronze Age Iberian individuals in the steppe ancestry turnover that took place over a period of four hundred years (roughly between 2400 and 2000 BCE), as mentioned in chapter 2, shows a lower steppe ancestry in the X chromosome than in the autosomes—17.3 versus 38.9 percent—thus suggesting a strongly male-mediated migration that likely took wives from local farming communities.[9] During the overlapping period it is still possible to find—especially in the south of Iberia—people with no steppe ancestry at all. But finally, around four thousand years ago, they disappear; everyone after this period of contact has steppe ancestry in around 20 percent of their genomes, and all men carry the Yamnaya-derived R1b lineage. At one site named Castillejo del Bonete (Ciudad Real, Spain), the forty-four-hundred-year-old double burial of a man and woman illustrates the nature of this interaction: the man has steppe ancestry (and carries the R1b chromosome) while the woman does not. From the isotopic analysis of her bones, it can be deduced that the woman originally had a maritime diet, meaning that she was not in the vicinity, as the site lies three hundred kilometers from the nearest coast. (We may well wonder if a double burial means that they both died simultaneously or if the widow had to follow her husband into the other world, but that's another story.)

One plausible scenario to interpret the Bronze Age genomic results is as follows: powerful migrating men had a large number of children, and their descendants constituted a new social elite with privileged access to resources (maybe triggered by the intergenerational transmission of livestock) that in turn enabled them to further reproduce and ensure the survival of their offspring. Strong organization in patrilineal clans can provide solidity to this emerging society over hundreds of years. The autosomal contribution of the newcomers halves with each upcoming generation, but not their Y chromosome, which can eventually become prevalent in the population (I will discuss this in detail in chapter 6). Under this premise, global genomic diversity would not have decreased. Also, though an untenable social dominance is not a requirement, strong social stratification with solid family ties certainly is. To be sure, this migration ushered in a dramatic level of

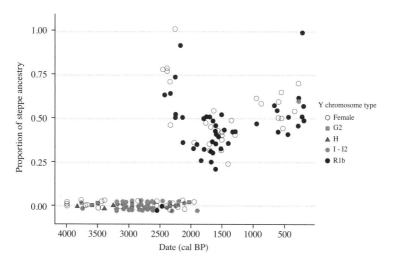

Figure 5.1
The male-biased arrival of the steppe ancestry in Iberia during the Bronze Age. Each dot represents an individual; filled circles are male (*gray*: Y chromosome lineages present during the Neolithic; *black*: the common Yamnaya lineage, R1b). A 40 percent genomic turnover can be estimated during this process, which took about four hundred years, with a 100 percent replacement in male lineage. *cal BP*: "calibrated Before Present" refers to a method used to correct for carbon-14 fluctuations over time; it yields more precise dating than raw radiocarbon dates. *Source*: I. Olalde, S. Mallick, N. Patterson, N. Rohland, V. Villalba-Mouco, M. Silva, K. Dulias, et al., "The Genomic History of the Iberian Peninsula over the Last 8,000 Years," *Science* 363 (2019): 1230–1234.

inequality, but society didn't collapse as one might imagine. Put simply, the wealthiest have reproduced more and with greater advantage, creating an enduring social elite, and the local women were incorporated into these clans.

The consequences of the extreme male bias in Bronze Age Europe can be seen in the uniparental markers too; while mitochondrial DNA diversity remained more or less the same from the previous Copper Age period, the Y chromosome underwent extreme changes in the frequency of some lineages. This trend is remarkable in the R1b, almost totally absent in the European Neolithic (so far it has only been described in a handful of dispersed individuals), and subsequently rising to frequencies of over 50 percent across the continent. In some cases, such as in Ireland or the Basque country in northern Iberia, currently above 80 percent of the males carry the Y chromosome lineage of the steppe pastoralists who entered Europe five thousand years ago. The previously common Copper Age Y chromosome lineages, such as I2, G2a, and H, were practically erased during the Bronze Age transition. In the north of Europe, the predominant Y chromosome associated with the steppe arrival is not R1b but rather R1a, which nowadays is also the prevalent lineage (for instance, in Germany). Why the steppe nomads are divided into these two paternal lineages is unclear; they could plausibly correspond to two different groups organized in their respective patrilineal clans. At the autosomal level at least, though, differences in the steppe ancestry component between northern and central-west Europeans are not discernible. Likely these were contemporaneous migrating groups from a common origin somewhere in the Pontic steppes. With our current data, it can be concluded that the arrival of steppe ancestry represented an extreme case of male-biased migration, not only in Europe, but across Eurasia.

Further evidence for the existence of higher female mobility across groups in this period comes from an analysis of the language. As we saw in chapter 2, steppe ancestry is plausibly related to the spread of the Indo-European languages, and lexical reconstructions of Proto-Indo-European shows a strong bias toward terms for the husband's relatives ("husband's mother," "husband's father," "husband's brother," "husband's sister," or "husband's brother's wife"), as opposed to an utter absence of those words for the wife's relatives.[10] In fact, the etymological interpretation of the word "husband's brother's wife" in Proto-Indo-European is equivalent to "traveler"; also, the synonymy of the verbs "to wed" and "to lead" suggests that brides were led away from their family homes to their husband's households.[11] These pieces of evidence are testimony to the unambiguous patrilocal character of the marriage institution among ancient Indo-European speakers.[12]

The spread of the steppe ancestry into South Asia showed some intriguing biases as well, at least in the available ancient samples genotyped. In Late Bronze to Iron Age individuals from the Swat Valley in present-day Pakistan, only 5 percent of the forty-four Y chromosomes analyzed were typical of the Central Asian steppe populations—all of them belonging to the well-known R1a lineage.[13] By contrast, the steppe ancestry was present with 20 percent frequency in the autosomes, indicating this ancestry was incorporated in the region mainly through females. In present-day South

Asians, however, the pattern observed is just the opposite, with a decrease of steppe ancestry in X chromosomes versus autosomes and a high presence of common Y chromosome steppe lineages, as we saw in the previous chapter. The strongest sign of this genetic component happens to be in Brahmins—members of the highest caste in Hinduism—and those who consider themselves to be of priestly status in modern-day India.[14]

It has been suggested that a similar scenario of male-biased migration might have happened in much later periods, such as the Anglo-Saxon transition in England during the fifth century CE. Despite the current estimates that the number of Anglo-Saxon migrants into Britain was quite low (from less than ten thousand to maybe two hundred thousand people), genetic studies reported high frequencies of Y chromosome lineages in modern Britain—ranging from between 24.4 percent in North Wales and 72.5 percent in central England—primarily deriving from southern Denmark, northern Germany, and Norway.[15] Computer simulations supported that these results could have been achieved with an apartheid-like social structure during the Anglo-Saxon period, based on differential reproductive success and limited intermarriage rates (especially between local British men and continental Saxon women, and not the opposite).[16] In these examples, it is perhaps paradoxical that although women are consistently the victims of inequality in a situation of conquest, conquered women seem to have had a much better chance of passing their genetic material to future generations than did their menfolk—a situation that we will see again in the Americas.

Such gender biases are not restricted to the Old World, though, nor to Indo-European speakers. Another case of sex-biased migration recently uncovered by paleogenomic studies took place in Remote Oceania after the disappearance of the Lapita culture about twenty-five hundred years ago. In the eyes of Westerners, the islands of Remote Oceania (a region that includes the archipelagoes of Vanuatu, New Caledonia, Fiji, Patau, Micronesia, and Polynesia) represent the stereotype of an earthly paradise, with their white sand beaches, pristine waters, and seemingly happy and easygoing way of life. In the eighteenth century, notions of the exotic beauty of their inhabitants and their willingness to engage in sporadic sexual intercourse took the popular fancy, thanks to the accounts of English sailors. We now know that Polynesian culture was actually quite complex, and that people there lived in hierarchical societies and had adapted to live in environments that could be remarkably hostile. Regarding promiscuity, what happened is that compared to Western morals—not to mention those of the first Christian missionaries—the range of acceptable behaviors was much wider and the social restrictions for sensual gratification less drastic. (This was probably still the case in the late nineteenth century, when the painter Paul Gauguin moved to Tahiti, and then met and married a thirteen-year-old native girl, all in the same afternoon.)

It might come as a surprise, then, that this Pacific paradise was the stage for yet another story of inequality for which paleogenetic evidence has only recently come to light. The inhabitants of Near Oceania, which comprises Australia, New Guinea, and the large Solomon and Bismarck archipelagoes (the last annexed by Germany in 1884 and named after Chancellor Otto von Bismarck), are genetically—and even phenotypically—very different from those of Remote Oceania. Part of their ancestry can be traced back to the arrival of modern humans in that region, about fifty thousand years ago, and they carry nearly 4.5 percent of Denisovan genes—an archaic human lineage from Asia.[17] Physically, they have a dark pigmentation as well as a robust skull with strong supraorbital bone reliefs, a prominent wide nose, and curly hair. The inhabitants of Remote Oceania have much lighter skin and display attenuated Asian-like traits, and their ancestry can be traced back to either Taiwan or the south of present-day China.

Genetic studies on the present-day inhabitants of the westernmost parts of Remote Oceania, such as Vanuatu, detected both ancestries several years ago: a Papuan-like component that ranges from 12 to 25 percent, and a Remote Oceania one.[18] As the former is clearly a more ancient substratum in the region (when first contacted by Westerners, the Papua New Guinean groups were in a Stone Age Neolithic way of life—prior to the discovery of metals—and thus constituted a unique case in modern humankind), geneticists believed the Papuan-like component was the original one in Vanuatu, which a latter ancestry of Asiatic origin might have overtaken during the Polynesian expansion that subsequently drove these intrepid seafarers to all the remote Pacific islands. This hypothesis made sense because the Polynesians were the first to develop the ability to sail long distances.

The first widespread archaeological horizon in Oceania is the Lapita culture, which expanded through Remote Oceania about 3,350 years ago, and produced a certain type of vessel decorated with geometric incisions and faced with obsidian. This culture is characterized by the use of domesticated animals such as dogs, pigs, and chickens, and the establishment of farming settlements based on the cultivation of yams, taro, coconuts, and bananas. It appears to have its origins in Taiwan or a nearby region on the continent, and spread in association with population movements triggered by the Neolithic expansion and improvements in navigational skills. Unmistakable evidence of Lapita presence first appeared in the Bismarck archipelago, moving from there to Tonga about 2,900 years ago and subsequently jumping over to Samoa. Its expansion ends there, however; Lapita people never reached the most distant corners of Remote Polynesia, and for about 1,800 years the Pacific saw no further colonization. The culture also ended abruptly about 2,500 years ago. It was only between 750 and 1,200 years ago that humans arrived on the furthest islands: Cook, Hawaii, Easter Island, and New Zealand. The Polynesian migration is the most extreme example of large-scale human movements; from the beginning of its Pacific expansion, close to Near Oceania, to Eastern Island, for instance, represents a span of thirteen thousand kilometers, most of them on the open sea.

A study led by Reich in 2016 retrieved genomic information from three individuals from Efate Island in Vanuatu, dated from three thousand years ago, and one skeleton from Tongatapu Island in Tonga, dated from four hundred years later. One of the Vanuatu remains was a skull separated from the rest of the skeleton and deposited in a Lapita pot. When Reich's team analyzed their ancestry, they discovered that these individuals showed no Papuan-like genetic traces, notwithstanding that group's supposed ancestry in this region. That is, the Papuan-like component turned out not to be the original one but rather superimposed atop the Polynesian-like ancestry after the Lapita period. Furthermore, the authors estimated the ratio of Polynesian ancestry in the X chromosome versus the autosomes and found the migration was also male biased.[19] But when did it happen?

In 2018, two more paleogenomic studies on ancient Vanuatu remains clarified the origins and influence of this unknown migration. In one, nineteen skeletons coming from Vanuatu, Tonga, Solomon, and French Polynesia were analyzed.[20] In the other, fourteen skeletons, again excavated in Vanuatu and dated from between 150 and 3,000 years ago, were genotyped.[21] Both studies demonstrated that the arrival of the Papuan-like component— likely from Bismarck—took place between 2,400 and 2,500 years ago, during a period that coincided with the end of the Lapita culture. In the skeletons from this critical period, Papuan-like ancestry was 100 percent. Not only had the general ancestry changed, the Y chromosome lineages, quite unlike those between Near and Remote Oceania, were modified (some Asian lineages, such as those from the O1 haplogroup, survived and have been found in two 2,000-year-old individuals from Epi island in Vanuatu). In Efate all paternal markers, however, such as K2 and M1, are typical of Papua. Nevertheless, the maternal counterpart, the mitochondrial lineage composition—mainly the Polynesian B4—remains unaltered throughout this migratory process. Again, the sex bias could be confirmed genetically; in a male individual from Malakula Island in Vanuatu, for instance, dated from between 2,320 and 2,690 years ago, the researchers confirmed a 50 percent excess of Remote Oceania-like component in the X chromosome (his Y lineage was typical Papuan M1, and his mitochondrial lineage was Polynesian B4).

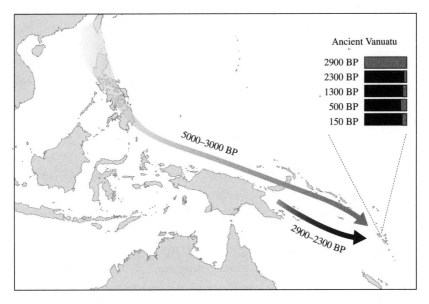

Figure 5.2
A sex-biased migration from Near to Remote Oceania based on paleogenetic analysis of remains from Vanuatu and other Polynesian archipelagoes. The proportion of the original ancestry (from 3,000 to 5,000 years ago) from mainland Asia or Taiwan over time is represented in gray; the subsequent ancestry (from 2,300 to 2,900 years ago) from New Guinea is represented in black. Interestingly, this major migration did not involve a language substitution. *Source*: M. Lipson, P. Skoglund, M. Spriggs, F. Valentin, S. Bedford, R. Shing, H. Buckley, et al., "Population Turnover in Remote Oceania Shortly after Initial Settlement," *Current Biology* 28 (2018): 1157–1165.e7.

In conclusion, the ancestry of some islands of Remote Oceania was determined by a Papuan migration superimposed on a previous layer. Why precisely groups less adapted to open sea travel followed this expansion remains a mystery. Yet in this case, interestingly, the original language was preserved. The inhabitants of Vanuatu kept the original languages, continuing to speak a language from the Austronesian family (it is estimated that there are more than a hundred different languages across these archipelagoes). In this case, for some reason—as we have seen on the Iberian Peninsula with the persistence of pre-Indo-European languages after the Bronze Age migration—they did not adopt a Papua New Guinean language, despite the genetic influence of the newcomers. One possibility that some researchers have suggested to explain this scenario is that they arrived in successive migratory movements rather than in one fell swoop. In such a long, slow process of modification in ancestry, it is much easier to maintain the original language. Others think that the Lapita vocabulary was best suited for a marine environment. This genetic transition nevertheless saw some language modifications, such as the adoption of a quinary system (that is, a number system in base five, which originates from the practice of counting with the fingers of one hand). Other cultural traits, such as penis sheaths, large nasal perforations, and cranial deformations by bandage, also point to influences from Near Oceania.

We have seen the power of paleogenetics for unraveling sex-biased contributions to ancient populations. But estimates can also be made by analyzing modern populations that underwent significant admixing in the last few centuries. The American hemisphere constitutes a paradigmatic example, as we saw in chapter 4.

Seldom in the recent history of humankind has a process of contact between different populations occurred on such a large scale as the discovery and conquest of the Americas by the Europeans. It involved the migration and blending of human beings from three continents: Native Americans, Europeans, and Africans.[22] The last were conveyed across the Atlantic as slaves from the early sixteenth century to the nineteenth century, when the trade was abolished. It has been estimated that the transatlantic slave trade involved the forceful transportation of about 9.6 to 15.5 million Africans; this is likely an underestimate as it is difficult to quantify those who died on the African continent itself or didn't survive the hardships of the Atlantic passage.

To give an idea of the magnitude of the American continental mixing, it has been estimated, for instance, that at the time of the "discovery" of Brazil by the Portuguese in 1500, there had been around 2.4 million Native Americans in that region, on whom almost half a million Europeans, primarily of Portuguese origin, were superimposed along with nearly 4 million African slaves. Clearly these three substrata could not interbreed equitably to form the current Brazilian population, and there is ample evidence that they did not.

The European colonizers, who traveled great distances to settle the American continent, were generally men of reproductive age. When they prospered in these territories it was certainly through imposition and force, to the detriment of the previous social structure. Populations that develop over a particular period after migration rarely involve random mating between locals and recent arrivals, as we have seen previously in this book; instead there arise patterns of social dominance that have an impact on subsequent generations. Often barriers are formed that are not equally permeated by both sexes, and this has discernible genetic consequences. In the Spanish colonies, all forms of interbreeding were perfectly defined, with the outcome being a caste system of the sort seen in India.[23] In Mexico, for example, a mestizo is the offspring of a male Spaniard and an indigenous woman; a *castizo* is the offspring of a male mestizo and a female Spaniard; a criollo is the offspring of a male castizo and a female Spaniard; a mulatto is the offspring of a male Spaniard and a female African; a *morisco* is the child of a male mulatto and a female Spaniard; a chino is the offspring of a male morisco and a female Spaniard; a *pelusa* is the offspring of a male chino and a female Amerindian; a *lobo* is the offspring of a male pelusa and a female mulatto; a *jíbaro* is the offspring of a male lobo and a female chino; an albino is the offspring of a male Spaniard and a female morisco, and so on. A cross between a male Spaniard and a female albino (one-eighth African) was known by the odd name of *saltapatrás* ("jump back") since this combination recovered a large share of European ancestry. These admixed children were the product of relationships between Spaniards and native women, mostly outside matrimony; for instance, in the mid-1500s in Guadalajara (Mexico), 85 percent of all children were born to unmarried parents.[24] These classifications did not survive the massive blending of the eighteenth century nor did they fully impede social mobility, but the origin of one's parents during the colonial period had a differential influence on the genesis of today's Latin American population. (In some sense, this system reflects the angst in sixteenth-century Castile about blood purity—that is, the absence in one's genealogy of Moorish or Jewish ancestry.)

The past two decades have seen numerous studies on the current populations of Latin America based on the mitochondrial DNA and Y chromosomes, from which a complex pattern of sexual asymmetries has begun to emerge. For example, in a study on Uruguayans of self-reported African origin, African maternal lineages accounted for 52 percent, but their male counterpart, the Y chromosome, was only present in 29 percent of the population. In addition, 29 percent of their mitochondrial DNA yet just 6 percent of their Y chromosomes were of Amerindian origin, while 19 percent of their mitochondria were European, as opposed to 64 percent of their Y chromosome lineage.[25] In Brazilians self-described as "white," 97 percent of their Y chromosomes were European, while just 39 percent of their mitochondrial lineages were (the rest, in almost equal proportions, were African and Amerindian). In one population from Colombia, 90 percent of the mitochondria were Amerindian, but only 1 percent of the Y chromosomes were of the same origin (94 percent were of European origin). Perhaps the most extreme case was a population in Curiepe (Venezuela), where 100 percent of the mitochondria were African, but there was not a single Y chromosome from that continent (80 percent were European and the rest Amerindian).[26] This bias is even more extreme if we consider that during the first centuries of the slave trade, about two-thirds of transported Africans were males. In a recent genetic study of twenty slave skeletal remains from remote Saint Helena Island (where Napoléon died in exile in 1821) in the tropical south Atlantic—about twenty-seven thousand "liberated" slaves were relocated there by Royal Navy ships intercepting slave ships between 1840 and 1862; seventeen individuals turned out to be males, and only three were females (none of them genealogically related), confirming the extreme male bias in the trade.[27] The existence of this initial bias in slave transport means that the African male contribution to subsequent American generations is extremely underrepresented.

Even before the availability of genomic data it was known that the current population of Latin America had been forged by means of brutal sexual asymmetries, and that European men had disproportionately contributed to future generations with regard to demographic figures to the detriment of Native Americans and African immigrants. To take reproductive priority

Español con India. Mestizo. 5

Mestizo con Española. Castizo. 6

Castizo con Española. Español. 7

Español con Negra. Mulato.

Mulato con Española. Morisco. 9

Morisco con Española. Chino. 10

Chino con India. Salta atras. 11

Salta atras con Mulata. Lobo. 12

Lobo con China. Gibaro. 13

Gibaro con Mulata. Albarazado 14

Albarazado con Negra. Canbujo. 15

Canbujo con India. Sanbaigo. 16

Sanbaigo con Loba. Calpamulato.

Calpamulato con Canbuja. Tente en el Aire.

Tente en el Aire con Mulata. Notentiendo.

Notentiendo con India. Tornaatras.

Figure 5.3

Painting comprising sixteen caste classifications in colonial Spain. The first one, mestizo, is the offspring of a male Spaniard with an indigenous woman. Preserved at the Museo Nacional del Virreinato, Tepotzotlán, Mexico. Image from Wikimedia Commons.

over these two populations, they monopolized the local female population to the point that "in a way, the Spanish conquest of the Americas was a conquest of women," as claimed historian Magnus Mörner in 1967.[28] To give but one illustration, in his chronicle on the conquest of Mexico, Bernal Díaz del Castillo mentions a certain soldier named Álvaro, born in Palos (Spain), who had thirty children in three years from various native women.[29] While this might be an extreme case, of the eleven children fathered by Hernán Cortés himself, three were born from three different native women (the remaining eight from three Spanish women). In general, Amerindian males, despite being native to the continent, had still contributed less than African males, especially in South America.

Another interesting aspect of this sort of study is the correlation between actual ancestry and self-reported identity, which has a sociological interpretation. For example, in Latin America, individuals with more than 5 percent Amerindian genes almost always considered themselves "Latino" (the term applied to a person from the Caribbean and Central or South America in the traditional census classification in the United States), even though this component was usually in the minority.[30] When the ancestry origin of this "Latino" group is examined, notable differences can be observed, with European percentages ranging from 56 percent in the Dominican Republic and 61 percent in Mexico to 84 percent in Cuba. In contrast, the Amerindian component ranges from 24 percent in Mexico to just 7 percent in the Dominican Republic; conversely, the African component is smaller in Mexico (3 percent) and much greater on that Caribbean island (28 percent).

In North America, a similar outcome from this blending process took place.[31] Looking at uniparental markers, a study found that African Americans from Philadelphia had 10 percent of mitochondrial DNA lineages and 31 percent of Y chromosome lineages of European origin, while European Americans from the same community had only 7 percent of mitochondrial lineages and less than 2 percent of Y chromosome lineages of non-European origin (the latter were almost exclusively Native American). Again, these data indicate that a gender-biased admixture took place in the American continent, with primarily European men and African or Native American women contributing to the ancestry of modern American populations.

Much more accurate ratios and microregional details can be obtained by studying whole genomes instead of uniparental markers to elucidate the dynamics of contact with Europeans, Native Americans, and Africans.

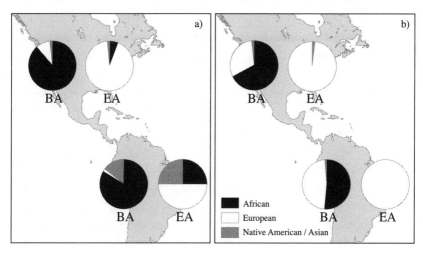

Figure 5.4
Contribution of different uniparental markers (*a*: mitochondrial DNA; *b*: Y chromosome) to two present-day American populations (Philadelphia in the United States, and Brazil in South America) with self-ethnicity attribution (*BA*: Black Americans; *EA*: European Americans). An extremely male-biased pattern of European origin can be seen. The Native American component is only significant along maternal lines and in South America, suggesting a different admixing pattern in Latin America compared with Anglo-America. *Source*: K. Stefflova, M. C. Dulik, A. A. Pai, A. H. Walker, C. M. Zeigler-Johnson, S. M. Gueye, R. G. Schurr, and T. R. Rebbeck, "Evaluation of Group Genetic Ancestry of Populations from Philadelphia and Dakar in the Context of Sex-Biased Admixture in the Americas," *PLoS One* 4 (2009): e7842.

In a study published in 2013 and directed by Stanford geneticist Carlos D. Bustamante, the genomic analysis of 55 current-day Puerto Ricans revealed that they carried nearly 12.5 percent of Amerindian ancestry. In contrast, the Amerindian mitochondrial DNA was around 60 percent, but not a single Y chromosome was of the same origin.[32] It must be remembered that on the arrival of the Spanish, many of the Caribbean islands were made up of peaceful agricultural communities, known as the Taino culture, which suffered an unprecedented demographic collapse in the subsequent decades. Despite commendable efforts by various groups to revive Taino culture and proclaim themselves descendants, Caribbean inhabitants able to trace more than 30 percent of their ancestry to the people of the American continent (and thus to some extent, to the original Tainos) are rare indeed. In 1508, an estimated 30,000 to 110,000 Tainos inhabited Puerto Rico alone; taking into account the preserved Amerindian percentage and the fact that there are currently 3.77 million Puerto Ricans, researchers have concluded that at least a large share of the ancestral genome can be reconstructed by simply analyzing and overlapping the scattered fragments that persist in current descendants—which remains an amazing application of the genomic techniques for reconstructing the great human family. Nevertheless, there have been complaints by some Taino associations that tried to correct the original publication, stating that this human group had not actually been wiped out. This could be true considering that complete ancestry is not a requirement for belonging to a cultural group, as is the case with many Native American groups of North America, for whom just one Amerindian mitochondrial DNA is required (even if the rest of the nuclear genome is almost entirely of European origin).

In a genomic study led by Joanna L. Mountain and published in 2015, the genetic ancestry of 5,269 self-described African Americans, 8,663 Latinos, and 148,789 European North Americans was analyzed. The authors discovered that the African Americans have on average an ancestry that is 73.2 percent African, 24 percent European, and 0.8 percent Amerindian. These values varied notably from one state to another; in some of the southern states, European mixing has been less significant due to social factors (interracial marriages have been legally forbidden until recently in some cases). The African Americans of Florida, Georgia, and Alabama, for example, have some 81 percent African genes, reaching a maximum of 83 percent in South Carolina (where the percentage of European genes reached only 15 percent). In states like West Virginia and Washington, African ancestry in African Americans was around 64 to 66 percent, with up to 34 percent of European contribution (more than doubling that in South Carolina). In contrast, European Americans display an overpowering ancestry of European origin, on average around 98.6 percent (with some 0.19 percent African and 0.18 percent Amerindian).[33] These values unambiguously indicate that the influx of African or Amerindian genes among the self-defined European group was socially deactivated (and legally sanctioned until recently). An additional factor to take into account is that until quite recently, US social and legal rules considered anyone who had a single African ancestor to be African American, however European they might have appeared outwardly, and however light their skin color. This is known as the "one-drop rule," and it came to replace the prescriptions of the US

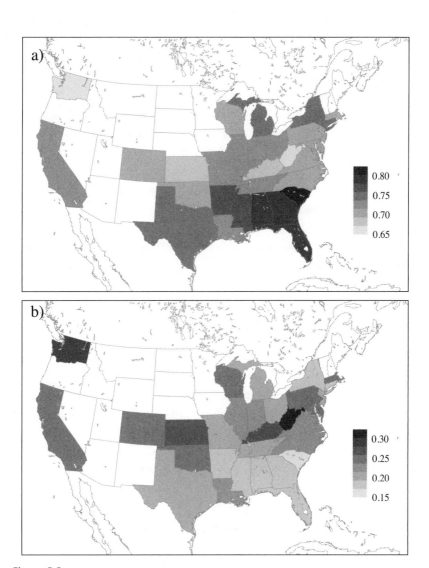

Figure 5.5
Distribution of ancestry of self-reported African Americans across the United States.
(*a*) Differences in levels of African ancestry in African Americans. (*b*) Differences in
levels of European ancestry in African Americans by state. States with limited data
(*white*) are excluded. The European component in the African American community
is much higher in the northern states. The differences are especially large in places
like South Carolina, Alabama, Arkansas, Georgia, Florida, and Mississippi. *Source*:
K. Bryc, E. Y. Durand, J. M. Macpherson, D. Reich, and J. L. Mountain, "The Genetic
Ancestry of African Americans, Latinos and Europeans Americans across the United
States," *American Journal of Human Genetics* 96 (2015): 37–53.

Census Bureau, which categorized African Americans as "pure" (unmixed), "mulatto" (of a European parent), "quadroon" (of a European grandparent), "octoroon" (of a European great-grandparent), and so on.[34] This rule would partially explain the genetic results since African ancestors in the European North American group have probably been "purged" from the group. In a study by geneticist Mark D. Shriver published in 2002, it was found that nearly 10 percent of self-defined African Americans actually had less than 50 percent African ancestry; from the genetic perspective, it would make more sense to consider them "white." As for Amerindian ancestry in Europeans, the figure peaks in western states like North Dakota, where it reaches 2 percent, even if it remains remarkably low in this community.[35]

Latinos (or Hispanics, as they are also known) in the United States have on average 18 percent Amerindian, 65.1 percent European, and 6.2 percent African ancestry. Even if they consider themselves descendants of the original inhabitants of the continent, their Native American ancestral category is actually the smallest component, not even reaching a fifth of the total genome on average, although higher values appear in the states bordering Mexico (for example, reaching 20 percent in New Mexico and 21 percent in Texas).

The differences in the ancestry of X chromosomes with respect to autosomes enables us to observe, for instance, that of the ancestors of African Americans, 5 percent correspond to European women while 19 percent correspond to European men. That is, the European contribution has a strong gender bias—almost four times greater through males than through females—due to the obvious issue of past social dominance. An interesting additional consequence of this potential gender bias through history is that women have contributed much more to humankind's genetic diversity than men.

What could possibly account for results like these on such a large scale? To appreciate how social mechanisms operate at the level of the individual, we have personal examples, in some cases of exceptional historical importance. Perhaps the best-known illustration is that of Thomas Jefferson (1743–1826), the third president of the United States and one of the drafters of the Declaration of Independence in 1776. Despite what we read in the second paragraph of that declaration (attributed to Jefferson himself)— "We hold these truths to be self-evident, that all men are created equal, that they are endowed by their Creator with certain inalienable rights, that among these are Life, Liberty and the Pursuit of Happiness"—he owned a plantation in Monticello, Virginia, that depended on slave labor for its operation. In the plantation books, we find that he possessed 163 slaves in 1795. Notwithstanding that he is described in the chronicles as a benevolent master and rarely resorted to the lash, the hypocrisy between his theoretical principles and reality strikes us as shocking—even if it was standard practice at that time. But there's more to the story. Jefferson appears to have fallen in love with one of his slaves, named Sally Hemings (who was actually just 25 percent African and 75 percent European). He was widowed at age thirty-nine, and when he moved to London five years later and then to Paris, Sally, who was just fourteen at the time, went with him as a domestic servant (actually, Sally and Jefferson's wife, born Martha Wayles Skelton, were half-sisters). All the evidence indicates that she became Jefferson's lover. She could have stayed in Paris as a free person, as slavery had just been abolished during the French Revolution, but pregnant with their first child, she preferred to go back to the United States and her master's family farm. The agreement, it appears, was that Jefferson would free her children (also his) when they turned twenty-one. As historian Annette Gordon-Reed states in her book about this relationship, "Though enslaved, Sally Hemings helped shape her life and the lives of her children, who got an almost 50-year head start on emancipation, escaping the system that had engulfed their ancestors and millions of others. Whatever we may feel about it today, this was important to her."[36]

In 1998, a genetic analysis of Jefferson's and Eston Hemings's descendants (Eston was Sally's youngest son) demonstrated that the two shared the same typically European Y chromosome lineage, indicating that they were quite probably related.[37] In the subsequent controversy, some critics said that they could not rule out the possibility that the father was not Jefferson but instead one of his relatives, such as his brother, Samuel, or cousin, Peter Carr (the latter hypothesis was subsequently discredited as Carr's Y chromosomes did not match). Still, historical evidence—the fact that all the births coincided with periods nine months after the president was in Monticello, for example, along with the gossip of the time and the physical resemblances—in combination with genetics indicates that Jefferson was in fact the father of Hemings's children.

The negative connotation of being African American could not possibly be more apparent than among the elite of the legitimate descendants of the third president of the United States. Counting nearly eight hundred among their ranks, they are members of the Monticello Association, which is comprised of genealogically confirmed descendants of Jefferson's daughters. It must have dropped like a bomb among this superexclusive club that some of Hemings's African American descendants requested admission. On May 5, 2002, they took a vote and decided to keep Hemings's progeny out of the club—a decision that in the words of Annette Gordon-Reed of Harvard Law School, was "scientifically unsound, historically unsound and morally bankrupt."[38]

We'll never know what they made of the implications of this affair, nor will we ever know if it was a love story or an abuse of power—or probably a bit of both. It may be hard to fathom that one of the country's founding fathers, a defender of liberty of Jefferson's stature, would abuse a slave girl of his. From today's ethical perspective, it's hard to discern such behavior as morally acceptable; however, some contemporaneous people may have thought otherwise given the morality of the time and given a society that still considered slavery normal. Still, Jefferson's story, shadowy as it is, gives us a glimpse of what the Americas' great melting pot must have been like, with its insuperable contradictions, its love stories, and certainly its share of suffering and domination. It is difficult to know how many such situations could be extrapolated; probably most would be attributable to rape and domination. Nevertheless, slave women like Hemings somehow managed to survive those terrible conditions for the sake of their children—just like numerous other mothers of the past and present.

Understanding such personal situations undoubtedly enables us to make sense of what has been observed at the population level throughout history. All these small personal decisions, one stacked atop another, form a composite picture through the years of the enormous gender inequalities that we've seen throughout this book. We cannot grasp one human dimension without coming to terms with the other.

6   In the Name of the Father

Your soul is indeed . . . a deep dark forest. But the trees in it are of a particular species—they are genealogical trees.

—Marcel Proust, *Pleasures and Days*

In the census of 1495, held for—of course—the purposes of taxation, Jaime de Lalueza was reported to have married María Lanao in the tiny Pyrenean village of Jaro, then in the kingdom of Aragon (now a region of Spain). A continuous line, eleven generations long (one every fifty years on average), connects that person and that surname to my own. Interestingly, throughout the first four hundred years, the Lalueza line moved no more than twenty kilometers south, reaching the equally tiny village of Abizanda, where my grandfather—the first to move to a big city, Barcelona—was born in 1878. In most Western societies, the Y chromosome is inherited in the same way as the surname (Iceland is particularly odd as the family name is shuffled each generation, with "son" being added to the father's given name if the offspring is a boy or "dottir" if it is a girl, effectively meaning that brothers and sisters have different family names." Therefore it is in principle possible to study the distribution of a particular Y chromosome lineage where surnames are uncommon enough that they likely have a single origin. While this is not the case with the most common surnames in Spain (in descending order, García, Rodríguez, González, Fernández, and López—all created multiple times, in many cases by adding the particle "ez," literally "son of," to a former common name), mine is clearly rare. The last time that I checked public phone directories I found only a handful of them across the world, almost all of them in Spain and belonging to people

I knew were related. Obviously we weren't carriers of a successful Y chromosome lineage—mine is, in fact, a rare T1a. But the question might be raised: What can make a specific paternal lineage expand well over its contemporaries? Besides the obvious answer of having sons instead of daughters, one possibility is to have a wealthy ancestor, likely in the distant past, who might have left not only numerous descendants but also the resources to feed them and their offspring for generations.

History is riddled with examples of legendary inequality (although there is usually much less information about the fate and number of their progeny). Croesus, who ruled Lydia—a kingdom of western Anatolia—between 560 and 546 BCE, has come down through history as synonymous with wealth (in England, the expression "richer than Croesus" is still used). He was the first to mint gold coins and donated a solid gold lion to the oracle of Delphi as well as two huge ceremonial kraters—one of gold, and the other of solid silver.

The Roman aristocrat and general Marcus Licinius Crassus often shows up on lists of the wealthiest people in history. At the moment of his death in the disastrous battle of Carrhae against the Partians in 53 BCE, his fortune was estimated (at current rates) to be some $170 billion. Crassus's initial wealth came from appropriating the farms of the victims of the proscriptions of Sulla, and was subsequently boosted through the sale of slaves, trade, and mining. But his notoriety for greed comes from the system that he organized to keep properties that had burned down in Rome. He had a group of five hundred builders under his command, and whenever a fire broke out—some of his contemporaries suggested that they themselves started them—the builders turned up to negotiate the acquisition at a discount of the damaged properties, which they then tore down, rebuilt, and resold, sometimes to the previous owners, at highly inflated prices. Roman moralists liked to tell the story that on Crassus's death at the hands of the Partians, they poured molten gold into his mouth. In more recent times lists of the wealthiest include not only autocrats like Czar Nicholas II of Russia but also businesspeople like Henry Ford, Andrew Carnegie, and John D. Rockefeller. The last insisted on the celestial origin of his wealth, famously stating, "God gave me my money. I believe the power to make money is a gift from God."[1] (Seven generations and more than a hundred years after the dynasty's founder amassed his wealth, the Rockefeller family still had a fortune estimated at $11 billion in 2016, according to *Forbes*.)

It seems evident that inequality within societies stems from the struggle for prestige, wealth, and power, which to a great extent is a contest for reproductive success. In unequal societies, especially in the past, those with a higher social standing had—or tried to have—more descendants than ordinary people. The Bible specifies that King Solomon had 700 wives and 300 concubines, and the chronicles tell us that Sasanian king Khosrow II had more than 3,000 young women in his harem (and a first wife who was the daughter of Byzantine emperor Maurice). In more recent times, the Imperial Harem of the Ottoman sultans occupied more than four hundred rooms at the Topkapi Palace, where hundreds of wives and concubines as well as numerous castrated slaves functioning as servants and guards lived.

Many of these women naturally produced offspring. Pharaoh Ramses II is said to have had more than 162 children, and Moroccan sultan Ismail Ibn Sharif (1672–1727) around 867. In recent times, the last Nizam of Hyderabad and one of the twentieth century's richest men, Mir Osman Ali Khan (1886–1967), had 149 recognized children, while the king of Swaziland, Sobhuza II (1899–1982), had no fewer than 210 (from 70 wives). We can think that these cases only happen in exotic cultures, but Mormon leader and founder of Salt Lake City Brigham Young (1801–1877), for instance, had 56 children from his 55 wives (early in 1846, he married 20 of them in less than four weeks, which I guess gives a different meaning to the idea of a honeymoon) that summed up more than 1,000 direct descendants in just a generation after his death and more than 30,000 at the beginning of this century.

Evidently, such extremely wealthy and powerful men, with their hundreds of descendants, were not in the mainstream of the population, which in current-day Western societies might generally produce one or two progeny per couple. Their genetic contribution to posterity is thus strongly skewed, and we can confidently say that for this same reason, they are more likely to have been our ancestors. Therefore the chances are greater that we're descended from kings than from contemporaneous peasants.

Still, we can easily work out that to some extent, we're all distantly related. The vast number of our ancestors, which doubles with each generation, means that we descend from nearly everyone who lived in the world more than a few thousand years ago. If, for example, we go back fourteen generations—about four hundred years ago—we find that we have more than 16,000 ancestors. Just six generations further and the number of genealogical ancestors rises to more than a million. This figure increases exponentially until it exceeds the number of people to have ever lived in the world. This is only possible because we share many of these potential genealogical ancestors not only with everyone else—including people in distant regions—but also with our own ancestors. In practical terms, it means that we have far fewer different ancestors than theoretically possible. In small, isolated populations, everyone who left descendants in the past is an ancestor of the entire current population.

At the individual level, having distant ancestors with higher reproductive fitness has little genetic impact. Although we inherit a copy of our 22 autosomes from our mother and another from our father, and this contribution halves each generation back (25 percent of our genome from each grandparent, 12.5 percent from each great-grandparent, etc.), the figures are not fixed, as they greatly depend on the points where chromosomes break each generation. That is, your mother, such as in chromosome 1, doesn't pass you her entire maternal or paternal copy (your maternal grandfather or maternal grandmother's copy) but rather a mosaic of both. The cellular process of creating this mosaic is a phenomenon called recombination that takes place during the production of sexual cells or gametes. Empirically it has been shown that you can inherit between less than 20 and more than 30 percent from each of your grandparents' genome, and the variation increases with each generation (this difference could explain why some grandchildren resemble one family branch more than the other).[2]

Women produce on average 45 chromosomal interchanges or recombination events when they make eggs, while men produce 26 chromosomal interchanges when they generate sperm cells; that is, for each generation, there are about 71 genomic recombination breaks.[3] If we project these figures toward the past, it implies that the number of ancestors who have contributed chromosomal fragments to our genome is much smaller than the number of genealogical ancestors. Our nuclear genome is composed of 46 chromosomes (44 autosomes plus 2 sexual chromosomes); going one generation before, our genome derives from about 117 chromosomal fragments (71 + 46) from our four grandparents. Moving another generation back, our genome carries 188 fragments (71 + 71 + 46) that derive from our eight great-grandparents. The fact, though, that the number of our ancestors doubles each generation back means that the number of chromosomal fragments that we've inherited from them soon falls off. For instance, ten generations back, we will have 1,024 theoretical ancestors, but only 756 fragments in our genomic mosaic, which means that some of our genealogical ancestors are not our genetic ancestors. The former are a random subset of the latter. (Hence the pride some genealogy enthusiasts feel when a famous ancestor is discovered in their family tree might not be warranted at all.)

It can be estimated with computer simulations that just six generations ago, some of your direct ancestors contributed nothing to your genome due to these fluctuations in genomic transmission.[4] Your entire genome is still there, scattered in bits across your 64 ancestors six generations ago; it is just that none of it derived from any particular individual. There is still, however, the possibility of inheriting a genetic marker from an individual who at some point vanished from the panel of genetic ancestors: if he is your great-great-great-great-grandfather through your father's line—and you are a male—you will get his Y chromosome in a direct line of succession.

The legitimacy of European royal families has been traditionally based on the bloodline, which is a reflection of Y chromosome descendants—with some remarkable exceptions in cases where women are the only surviving children, such as Queen Elizabeth I (and the second, of course). Consequently, we can reconstruct some pedigrees and assume that they share the same Y chromosome lineage (although potential extrapair paternities can certainly give the lie to this assumption).

The most important royal family in Europe has lasted more than a thousand years and derives from Hugh Capet (born around 939 and died in 996). It produced 36 kings and queens of France, 9 of Portugal (plus 20 more through an illegitimate branch), 11 of Naples, 3 Latin emperors of Constantine, 4 of Sicily, 2 of Etruria, 16 of Navarra, 4 of Poland, 2 of Albania, 11 of Spain, 4 of Hungary, 2 of the Kingdom of the Two Sicilies, and 2 of Brazil as well as innumerable dukes, counts, and marquises. It has been estimated that today there are around 6,500 living descendants of Capet, although only 2 of them are actually heads of state: Henri, Grand Duke of Luxembourg, and Felipe VI, king of Spain.[5]

The need to maintain royal hereditary lines isolated from the rest of society probably modified some genetic variants in those families substantially—increasing or decreasing them with respect to the allelic frequencies of the main population. In some cases, the accumulated consanguinity had negative effects after several generations. The Spanish Habsburgs are a famous example. Their last king, Charles II (1661–1700), also known by the rather deplorable nickname "El Hechizado" (the Bewitched), is estimated to have had a consanguinity index of 25 percent, which would be the equivalent of being the offspring of a brother and sister. This astounding ratio means that for a quarter of his genome, the paternal and maternal copies of each autosome would be identical by being derived from a common ancestor. It has also been estimated that 95.3 percent of Charles II's genome would derive from only 5 ancestors. This is a consequence of the fact that since 1550, no outsider married any member of the Spanish royal family. The resulting situation can be easily explained by checking some of his most recent ancestors: the father of the king, Phillip IV, was the uncle of his mother, Marianne of Austria; his great grandfather, Phillip II, was also the uncle of his great-grandmother, Ana of Austria; and his grandmother, Mary Ana of Austria, was at the same time his aunt because she was sister to Phillip IV too. Instead of having 32 great-great-great-grandparents like anyone without consanguinity in their family, Charles II had only 14! After his death in 1700, grim details of his autopsy were leaked to the court by the marquis of Ariberti: "[The corpse] did not contain a single drop of blood; his heart appeared of the size of a peppercorn; the lungs, corroded; the intestines, rotten and gangrenous; he had a single testicle, black as coal and his head was full of water."[6]

As a side effect of this accumulated lack of genomic variation—which likely exposed various deleterious mutations that would go unnoticed in heterozygosity—the unfortunate Spanish king had a long list of physical and cognitive problems, among them serious difficulties in speaking (he could not utter a word until he was four years old) and eating, weakness in his legs (he was unable to walk until he was eight), liquid retention, eye problems, frequent fevers and nausea, gastrointestinal disorders, sterility, and a certain degree of intellectual disability. All these conditions ultimately led to his death without heirs. This fact effectively meant the extinction of the House of Habsburg, founded in the eleventh century by Radbot, count of Klettgau, who built the Habsburg Castle that gave the royal house its name.

We have already seen that these long pedigrees mean nothing in terms of genetic inheritance; although Queen Elizabeth II of the United Kingdom can trace her ancestry back to distant kings such as William the Conqueror, their genetic contribution to her actual genome is a much messier issue, mainly because royals are often intermingled over time with the same ancestors, increasing the endogamy of these family lines. According to genealogy programmer Michael Ruby, Queen Elizabeth II descends from William the Conqueror through at least 752,409 family paths; the pair of royals are separated by 31 generations, which means William is only one of her 2,147,483,648 ancestors from that time. He is also a genetic ancestor of at least 191,177 others of these ancestors, however, and over a million more at more recent branches—who in turn descend from William via different connections—of her family tree. In addition, there are another 353,057 ancestors deriving from William's sister Adelaide (who of course shared about half her genome with the king). The conclusion is that Queen Elizabeth might have about 0.076 percent of William's original genome on her own, although this is likely an underestimate because we have to assume that many royal genes "infused" to the commoner's genomic landscape across centuries through illegitimate descendants. If we take the first value—the real one being impossible to estimate—there's a good chance that the queen currently carries none of the conqueror's genes.

A 2007 genetic study revealed that 7 English people from Yorkshire shared a common ancestor in the seventeenth century and a Y chromosome originally from sub-Saharan Africa.[7] As many as 300 people of African ancestry were known to be present in the courts of Henry VII, Henry VIII, and Elizabeth I during the Tudor era.[8] Documents from the early sixteenth century mention that John Blanke, who was nicknamed "the Black Trumpet" and played in jousting tournaments, was employed by Henry VII (it has been suggested that he came to England as an attendant of Catherine of Aragon in 1501). Blanke enjoyed some celebrity at court and even married a London woman. Yet four centuries of subsequent marriages with English people might have totally erased the African ancestry from the genome of these Yorkshire descendants. The fact they had an uninterrupted paternal succession during so many generations is in itself remarkable, especially if we consider that most noble houses from the Tudor era have already vanished for the same reason. Again, a potential discrepancy between the Y chromosome lineage and an overwhelmingly "English" autosomal genome could explain these results.

Nevertheless, a succession of fathers who in turn had many more children than contemporaneous males, coupled with a continuation of such a trend for several generations, might clearly produce a discernible genetic pattern, at least at the Y chromosome level. Some examples of this turn up in recent history, but at least one seems to have its origin in a historical character: Genghis Khan. There is a disputed quote attributed to him (also used by Arnold Schwarzenegger in the 1982 epic film *Conan the Barbarian*): "The greatest happiness is to vanquish your enemies, to chase them before you, to rob them of their wealth, to see those dear to them bathed in tears, to clasp their wives and daughters to your bosom." The final words, even if he never pronounced them, are revealing of the Mongol view of life. An additional reward to any violent conquest in ancient times was the possibility of having access to more women, and a secondary consequence was that conquerors could have more offspring than conquered peoples.

The personal story of Temüjin (ca. 1162–1227), later known as Genghis Khan, is astonishing. The nomadic chieftain who emerged from the Mongolian steppes, having been brought up in appallingly adverse conditions, did not learn to write and never saw a city, yet he created from scratch the largest empire in human history (his title, Genghis Kha Khan, meant "emperor of all men"). According to the anonymous account *The Secret History of the Mongols*, at the premature death of his father, Yesukai, when Temüjin was still in his teens, rival clans fell on the possessions of his own clan, which had previously held certain rights to northern Gobi territories. Temüjin had to fight for his life as well as save those of his mother and six brothers. In one instance, he escaped from the campsite of some Taidjut enemies who had captured him as he hauled a heavy *kang* (a wooden beam placed on the shoulders with the hands tied at one end). On another occasion, he was left for dead in the snow with an arrow piercing his chest.[9]

Over time Temüjin forged alliances with different clans, and surrounded himself with devoted followers who were skillful warriors and notable commanders too. His prestige, based on his proven charm, bravery, and wisdom, increased among the Central Asian tribes. In turn, he was always generous with those who served him; probably the trait he valued most was loyalty.[10] Around the year 1203, Toghrul, also known as Wang Khan, chief of the Keraites, betrayed Temüjin and tried to kill him. This action led to a war between different Mongol tribes from which Temüjin's forces emerged victorious. In 1206, a meeting of the clans at the Onon riverbank determined the union of all tribes under his leadership and his adoption of the title Genghis Khan. This unification represents a turning point, and the start of the Mongol expansion to the east and west.[11]

A recent study has estimated that the Mongols killed about forty million people during the subsequent expansion. This devastation led to the abandonment of vast extensions of arable lands that were subsequently recolonized by forests, which in turn absorbed some seven hundred million tons of carbon dioxide—thus effectively modifying the planet's climate.[12] For Christians, Genghis Khan was the incarnation of the Antichrist, and his name was associated with savagery. Muslims also considered the Mongols the scourge of God (the sacking of Baghdad in 1258, which caused two million deaths according to contemporary chroniclers, ranks among the worst examples of a city's destruction in history). When Genghis Khan died in 1227, his son Ögedei inherited an enormous empire extending from Armenia to China, and continued to wage conquests in Asia and Europe; in the latter, the Mongols ravaged the kingdoms of Moravia, Poland, and Hungary, and even laid siege to Vienna.

Figure 6.1
Portrait of Genghis Khan, by an anonymous painter of the Yuan dynasty (1279–1368). No contemporary portraits of the founder of the Mongol Empire have survived. Image from Wikimedia Commons.

Notwithstanding their negative image, the Mongols were permissive with regard to all religions and provided a law code—the Yassa—to more than fifty Central Asian populations. Some researchers argue that they built the modern world as we know it; by uniting Eurasia into a single empire, they enabled the circulation of all kinds of technological innovations throughout this vast territory, including paper, gunpowder, paper currency, and the compass. Moreover, the empire created the core of a global system by allowing free trade, the circulation of knowledge, the disconnection of politics from religion, and diplomatic immunity, among other modern phenomena.[13]

The personality of Genghis Khan remains elusive for modern historians; the Mongols did not write chronicles. In fact the chroniclers of the Mongol conqueror were his enemies, those who from their cities saw the arrival of horse-riding hordes seemingly bent on looting and destruction. Quite often Genghis Khan has been compared to Alexander the Great, but one of the most obvious differences between the two are their legacies. At the death of the Macedonian king, his generals disputed the conquered territories in a long series of wars. Alexander was expecting a son—who was not to be born—and had a newborn child—never to grow to adulthood—from two different women. In contrast, Genghis Khan's third son, Ögedei, ascended to supreme khan and kept the Mongol Empire stable for most of his reign. A grandson of Genghis Khan, Kubilai Khan, continued to dominate vast territories and founded the Yuan dynasty in China, which lasted until 1368 before being replaced by the Ming dynasty. The son of Hulagu, Kubilai's brother, dominated Persia, and other descendants reigned over China and regions of Central and South Asia for centuries. In the end, the incredibly long distances separating one khan from another led to the fragmentation of the Mongol Empire and its eventual disappearance.

But clearly there were quite a few generations of reproductive dominance by Genghis Khan's successors in the Mongol territories. All of them—including the founder of the dynasty himself—had many wives and in some cases large harems. It is not inconceivable that they spawned thousands of descendants all over the earth who would carry the same Y chromosome as Genghis Khan—although progressively less of its nuclear genome, as we have seen previously in this chapter. To leave descendants, one must not only have many offspring for many generations but also enough resources to ensure that a significant number of them reach adulthood and have offspring in turn. We have seen in previous chapters that infant mortality was a prevalent cause of death in the past; factors such as good nutrition would be a differential trait for offspring survival in hierarchical societies.

In a work published in 2003 led by geneticist Chris Tyler-Smith, researchers described the most abundant Y chromosome lineages in Asia and found that one particular haplotype (a combination of genetic markers inherited as a single block because most of the Y chromosome does not recombine) was so common that it was carried by around sixteen million males in Asia—that is, about 0.5 percent of the world's men. When the analysis was circumscribed to the geographic extension of the Mongol Empire, that lineage represented approximately 8 percent of all men in that region. And when they estimated its age, based on the accumulated additional mutations in the core haplotype and using a widely accepted mutation rate over twenty-five years of generation time, they found that the lineage emerged and expanded about seven to thirteen hundred years ago, in a period that overlapped with the Mongol Empire.[14]

Naturally, over the course of time, not all descendants carry exactly the same combination of genetic markers across the Y chromosome. With successive generations, new mutations appear in various descendants, creating a pattern with many short branches spanning from a central node that is the original combination of genetic variants (or haplotype) defining the founder's lineage. The authors named this pattern "starlike," and in fact it can be observed in diverse species and different genetic markers, whenever a population or species experiences demographic growth in a short period of time. The more recent the starlike expansion, the clearer the pattern can be observed because additional demographic events will tend to create subsequent stars and secondary branches that can "erase" the original ones from the common ancestor.

A distribution of this type cannot be generated by chance and has to be shaped by some kind of selective force. The researchers estimated that Genghis Khan's lineage grew with the equivalent of a selective coefficient (a figure that measures the selective advantage in terms of the differential fitness of a specific mutation, or combination of genetic variants, over the rest) of 1.36 per generation (if one individual has the population average of offspring, their selective coefficient would be 1). This is a remarkable selective force, similar to the one estimated for the spread of a mutation that turned fair moths black during Victorian times because they were concealed

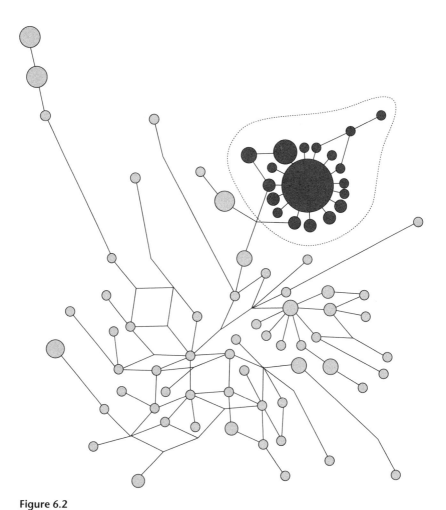

Figure 6.2
Phylogenetic tree of Asian Y chromosomes showing the starlike structure of the lineage attributed to Genghis Khan. The size of the circles is proportional to the number of individuals genotyped. Each line represents a mutation apart. A starlike shape involves a rapid expansion from the original lineage (*central node*) that can be dated with the application of the estimated Y chromosome mutation rate. *Source*: T. Zerjal, Y. Xue, G. Bertorelle, R. S. Wells, W. Bao, S. Zhu, R. Qamar, et al., "The Genetic Legacy of the Mongols," *American Journal of Human Genetics* 72 (2003): 717–721.

to bird predators on the darkened trees of the Industrial Revolution. There are few genes in the Y chromosome that could explain such a biological advantage, though, and the most plausible explanation was the existence of social dominance and privileged access to women by one man, or a group of paternally related men. The authors concluded that the genetic and historical evidence to explain this striking pattern pointed to Genghis Khan and his offspring. The potential founder of the Y chromosome haplotype would be placed about thirty generations ago, and an increased reproductive capacity, maintained over several generations, would suffice to explain the pattern described. Their interpretation is supported by the estimated age of the founder and the geographic distribution as well as some additional evidence. For instance, the Hazara from Afghanistan consider themselves descendants of the Mongols who fled Persia after the collapse of Mongol dominance there, and in fact they have the presumed Genghis Khan haplotype in high frequencies, even if it is almost absent in neighboring populations in Afghanistan itself or Pakistan and India.

Nevertheless, as we have previously seen, being the "grandfather" of all these people across Asia is rather irrelevant from an autosomal point of view (that is, for the rest of the genome). Many descendants will likely carry nothing at all from the original Mongol conqueror or at most his ancestry will appear in about 4 percent of their genes, according to the researchers' estimates.[15] Or maybe even less, according to other researchers, mainly Chinese academics; subsequent analyses have suggested that the lineage attributed to Genghis Khan was maybe prevalent among ordinary Mongols.[16] While this could be the case, the explosive increase of this Y chromosome still holds, and Genghis Khan is still a plausible explanation for the pattern observed.

Of course, even among the known descendants of a particular genealogy we can expect to find some males who do not match the expected Y chromosome lineage either because of adoptions or extrapair paternities arising through the generations. This could happen even in royal or aristocratic families. For instance, in the genetic study undertaken to identify the possible remains of King Richard III—the same one who, according to Shakespeare, yelled out, "A horse, a horse! My kingdom for a horse!"—found underneath a parking lot in Leicester in 2012, the researchers selected some living descendants of the king's ancestors to compare their Y chromosome to that retrieved from the skeletal remains (the only legitimate son of Richard III, Eduard of Middleham, died in infancy in 1484) and found discrepancies in some of them. The skeletal signs of scoliosis in the spine as well as violent injuries produced by swords and probably a halberd matched

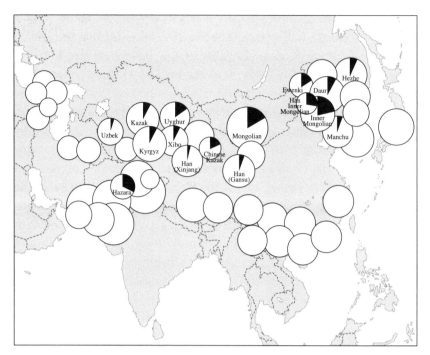

Figure 6.3
Distribution across Asia of the Y chromosome lineage attributed to Genghis Khan. The size of the circles is proportional to the number of individuals. The proportion of the putative Genghis Khan lineage is represented in black. With the exception of the Hazara (Pakistan), the lineage is present only across Central Asia and northern China. Source *Source*: T. Zerjal, Y. Xue, G. Bertorelle, R. S. Wells, W. Bao, S. Zhu, R. Qamar, et al., "The Genetic Legacy of the Mongols," *American Journal of Human Genetics* 72 (2003): 717–721.

with the historical accounts of the king's appearance—exaggerated by Shakespeare in his play—and the circumstances of his death at the battle of Bosworth.[17] The researchers investigated five living descendants of the fifth Duke of Beaufort (1744–1803), who was in turn a descendant of Edward III (1312–1377)—great-great-grandfather to Richard III—through John of Gant (1340–1399), first duke of Lancaster. None of the five living relatives had the same Y chromosome as that of the putative remains of Richard III—thereby suggesting an extrapair paternity along the paternal line—and one of them even had a different haplotype—thus indicating yet another extrapair paternity event in the last two hundred years. If the remains found indeed belong to the king and there was an extrapair paternity in the English royal family, the consequences are interesting: if it took place between the reigns of Edward III and John of Gant, then all kings descending from the latter—Henry IV, Henry V, and Henry VI—would be delegitimized for the throne—if this indeed depends exclusively on the bloodline. There were malicious rumors when John of Gant was alive that his father was in fact a Dutch butcher; if there was a false paternity there, or between John of Gant and his son, John Beaufort, this would invalidate the whole Tudor royal branch because the mother of Henry VII was the daughter of John Beaufort. If it happened between the reigns of Edward III and Richard III himself, then paradoxically the king who killed the two sons of his brother for presumed illegitimacy—the famous princes of the Tower—would have in turn been deemed illegitimate to reign.

It is difficult to know how many extrapair paternity events are expected to be found along paternal lines. In a recent study attempting to address this question, researchers from Leuven University and collaborators reconstructed deep genealogies from parish records spanning in some cases five hundred years and subsequently genotyped two living relatives descending from a common ancestor but being in different family branches. They found that extrapair paternities in married couples varied by more than an order of magnitude, between 0.4 and 5.9 percent, peaking among families of low socioeconomic status living in densely populated urban areas during the Industrial Revolution.[18] Although of course it is difficult to know how these results can be extrapolated to different periods and cultures—considering that the social context was clearly influential for determining extrapair paternities—it seems clear that this is a factor to be taken into account for disrupting Y chromosome lineages, sometimes reshaping the ongoing starlike phylogeny along specific subbranches.

The Y chromosome of Genghis Khan is an extreme example, but not the only one. The phylogeny of the paternal chromosome, described in different branches identified by letters of the alphabet, shows here and there other starlike configurations that correspond to other, more local expansions. These likely replicate the Mongol example on a smaller geographic scale. Another well-known instance is that of Niall of the Nine Hostages in Ireland. Niall was an obscure king and funder of the Uí Néill dynasty that dominated the northern part of Ireland from the sixth to tenth century. The chronicles say that he lived between the end of the fourth century and the beginning of the fifth, although there is no solid evidence of his existence, and all we know about him can be regarded as legendary. What interests us, however, is that he had two wives: Inne—from whom he had a son, Fiachu—and Rignac, who gave him seven more sons, with names that might sound elfish to our ears: Lóegaire, Éndae, Maine, Eógan, Conall Gulban, Conall Cremthainne, and Coirpre. These sons gave rise to several dynasties of Irish and Scottish clans, and even if the origins are not well documented, it seems reasonable that they carried the same Y chromosome lineage and likely had a reproductive advantage over contemporaneous Irish males. In a study on the Irish Y chromosome gene pool directed by geneticist Dan Bradley in 2006, the researchers discovered that nearly two to three million present-day men (mostly in the north of Ireland, but also in Scotland and the United States) derive their Y chromosome from a common ancestor who lived around the year 500 CE. This lineage was present in about 21 percent of the men from the north and northwest of Ireland, and although it is almost absent in the south, it accounts for 8 percent of all the island's inhabitants.[19] Precisely because of the coincidence in time and the potential social superiority of his descendants, this observation could be associated with Niall and his sons even though this identification is, as in the case of Genghis Khan, speculative.

Another study revealed a starlike haplotype of C3 Y chromosome lineage that was especially prevalent in northeastern China (ancient Manchuria) and Mongolia. The dating of the original lineage from the observed modern diversity yielded an estimate of six hundred years, which coincides with the life of Giocangga (who died in 1582), the grandfather of Nurhaci (1559–1626).[20] The latter unified and reigned over the Manchurian tribes, and his descendants conquered China and established the Qing dynasty—which replaced the previous Ming dynasty—lasting from 1644 to 1912. The aristocratic class of this dynasty was comprised of descendants of Nurhaci who possessed enormous economic resources; it has been estimated that at the beginning of the twentieth century, this elite—which likely shared the Y chromosome haplotype—was officially represented by more than eighty thousand people. Undoubtedly their privileges could boost the distribution of their Y chromosome that nowadays includes about 3.3 percent of all East Asian men.

Also in China, the sequencing of a large sample of Y chromosomes revealed three subbranches of the O3a-M324 lineage with clear starlike

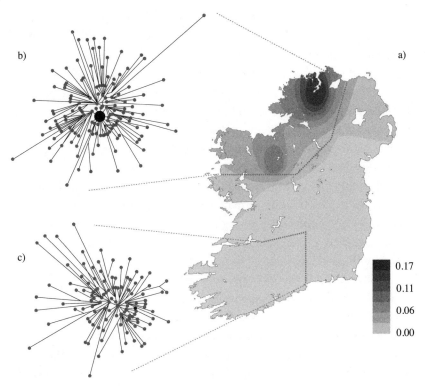

Figure 6.4
Phylogenetic tree of Irish Y chromosomes showing the starlike structure of the lineage of Niall of the Nine Hostages. The size of each circle is proportional to the
number of individuals, and each line represents a mutation apart. The lineage of
Niall of the Nine Hostages (*black*) is predominant in the north of Ireland (where it is
currently present in three million Irish males, attaining a frequency of 17 percent),
but almost absent in the south. *Source*: L. T. Moore, B. McEvoy, E. Cape, K. Simms,
and D. Bradley, "A Y-Chromosome Signature of Hegemony in Gaelic Ireland," *American Journal of Human Genetics* 78 (2006): 334–338.

phylogenies that dated from the Neolithic period, six thousand years ago.
Nearly 40 percent of Chinese present-day Y chromosomes derive from
these three paternal founders, who researchers named "Neolithic super-
grandfathers." The origin of these lineages seemed to be located in the Yel-
low River region, and the researchers speculated whether two of them could
be the legendary emperors Yan and Huang, considered to be the ancestors
of the Han Chinese.[21]

A global survey of human Y chromosome lineages led by geneticist Toomas Kivisild recently concluded that many of those that are the most common in present-day populations were rare in the past and that their current predominance took place in short periods of time.[22] This and subsequent studies found an abrupt bottleneck in the diversity of the paternal chromosome about five to seven thousand years ago—coincident with the end of the Neolithic era—and a subsequent explosive expansion of some lineages—coincident with the Bronze Age period. This trend could only be attributed to differential reproductive success in some males and can be seen in all Old World regions (although it is perhaps less pronounced in Africa). Still, the increase in male reproductive variance alone does not seem enough to explain the strong bottlenecks observed. From an evolutionary perspective, the explanation seems to be related to different groups organized in patrilineal clans that competed among themselves by leaving more descendants, in parallel with the world's growing population.[23] Those expansive Y chromosome branches include the well-known Yamnaya R1b and R1a lineages that started their expansion in the Bronze Age. The typical starlike shape can be seen in these branches, albeit in a much more complex form and with numerous ramifications, probably formed by the antiquity of the original lineages and the scale of their dramatic contribution to subsequent generations.

The most prevalent lineages nowadays were rare prior to becoming the majority; as a consequence, this process of expansion provoked a loss of global genetic diversity due to the decrease or even disappearance of alternative Y chromosome branches, mainly during the Bronze Age. Throughout history, we can affirm that no other genetic marker has been more affected by past inequality than the male chromosome. This also means that many males nowadays share an ancestor in that fraction of their genomes that derives from around five thousand years ago. In contrast, the maternal counterpart, the mitochondrial DNA, does not show any such bottleneck, thereby suggesting that the female population was much more stable than the male one, and increased continuously over time, especially around ten thousand years ago with the advent of agriculture. Therefore what I am describing here echoes the existence of substantially different population dynamics between males and females during the last thousands of years.

The starlike formations interest people because they are visual and intuitive, and enable us to understand quick expansions of specific haplotypes

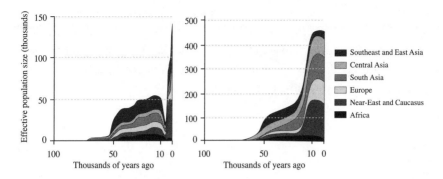

Figure 6.5
Cumulative genetic diversity of the Y chromosome (*left*) and mitochondrial DNA (*right*) lineages with their estimated divergence time by geographic area. The population size is the size of an ideal population based on genetic parameters. The paternal chromosome underwent a severe bottleneck around the Bronze Age on every continent except Africa. mtDNA experienced a continuous increase with no bottlenecks over the last tens of thousands of years. *Source*: M. Karmin, L. Saag, M. Vicente, M. A. W. Sayres, M. Järve, U. G. Talas, S. Rootsi, et al., "A Recent Bottleneck of Y Chromosome Diversity Coincides with a Global Change in Culture," *Genome Research* 25 (2015): 459–466.

at a glance. For geneticists, these patterns shed light on past cases of social superdominance that can even be dated. We are likely seeing the genetic footprint of powerful people in historical times, but also anonymous people in prehistory who for some reason were reproductively successful prior to written records. In some cases, we will never know their names or circumstances, but we are still able to see the signs of their reproductive dominance—again involving strong control over contemporaneous women.

The more starlike formations that we can find in a specific period, the more inequality would have existed in that moment from the past. Some of these episodes can be quite restricted geographically, but others would have left a disproportionate mark—considering that they started in one man at some point—in humankind. This means that individual people might have significantly modeled modern human populations while also implying that contemporaneous people were prevented in one way or another from contributing to subsequent generations, at least on the same scale.

The achievements of powerful people in the past have been traditionally recorded by chroniclers and historians, but now in some sense they can be explored by geneticists too. And genomics is making different individuals of each of us. The comprehension of this process of past inequality will likely serve to connect two genetic spheres that have usually been kept separate: the individual one—which we might call "microhistory"—and the population one—traditional "history."[24]

7   The Future of Inequality

Eugene suffered under a different burden—the burden of perfection.

—*Gattaca*

I am revising this book during a trip to Montenegro, in the Balkans, at the dawn of the second COVID-19 outbreak of 2020. With the help of drones, local archaeologists have located 210 tumuli from the Bronze Age, almost all of them unexcavated, in a small region between Tivat and Budva, overlooking the Bay of Kotor. Considering the complicated terrain of this area, with mountains up to two thousand meters high but close to the sea, archaeologists speculate that they could correspond to pastoralist groups seeking pasturelands, perhaps associated with the spread of the steppe nomad ancestry. This hypothesis remains yet to be tested, but this finding shows that the genetic cases I described in this book and their relationship with past inequality will continue to accumulate in the coming years, shedding new light on different regions and periods. As we have seen, we just need to dig up the evidence.

But this year 2020, unsettled by the COVID-19 pandemic, is bearing an unprecedented negative impact on the economy that will disproportionally affect lower-income workers, thus likely increasing inequality around the world.[1] Once the current coronavirus pandemic is over, we'll discover what impact inequality had on people's survival.[2] Ordinary people have had no choice but to keep working in unsafe conditions at the risk of being more exposed to the coronavirus; they're also more likely to have health conditions associated with poverty, such as asthma and diabetes, which represent a higher risk of dying from this disease. Economic disparities, to some extent ethnically structured, influence poor health outcomes in countries like Great Britain and the United States.[3] In New York, for example, the Latino and Black American communities have been hit harder than white ones. Poorer zip codes across the city, where people live in small, crowded apartments, and cannot work from home or fly to second homes in the bucolic countryside, are where the highest death rates are concentrated. Even in states with low population densities, Black Americans are disproportionately represented; in Maine, for instance, Blacks represent about 21 percent of the infected people even though they make up just 1.4 percent of the population.[4]

A paramount illustration is what happened in Singapore at the beginning of the pandemic. For weeks, Singapore implemented a highly technological lockdown, with a smartphone tracing system, and recorded fewer infections—and was hailed as an example to the world—only to see a major outbreak several weeks later. The second coronavirus wave, in mid-April 2020, surged in cramped migrant facilities on the city's outskirts, where often ten to twenty people shared dormitories and technological monitoring wasn't viable. In short, the dream of technology came up against the world's reality.[5]

When vulnerable social circumstances coincide with specific communities, the pandemic crisis can have obvious implications for the genetic diversity of those groups. In Los Angeles, for instance, people of Polynesian descent, including native Hawaiians, experienced an infection rate six times higher than those of their white neighbors. And the Navajo Nation reported in the first half of 2020 more per capita cases of COVID-19 than any region in the United States with the exception of New York and New Jersey. The combination of poverty, a poor public health system, and a high prevalence of diabetes—a known risk factor for the coronavirus—could explain these figures on the Navajo reservation.[6] Under the current scenario, it's plausible this pandemic will represent a certain genetic bottleneck in some indigenous communities across the US continent of a magnitude comparable to previous pandemics. (The 1918 "Spanish" flu, for example, killed an estimated 24 percent of the Navajo population.)[7]

Therefore this book, despite what it seems, is not just a study of the past. More than ever, we need to explore past episodes of increases in income disparity to understand the world in which we live today. To paraphrase philosopher Søren Kierkegaard, inequality must be lived forward, but can only be understood backward. This is what I've done in these pages: looked into the human past with the genetic tools we have in our possession to study how common inequality was in the past and how this has contributed to modern human diversity. The emerging picture is that it was so prevalent that we can confidently say we are the product of thousands of years of genomic inequality. We carry in our genomes a significant share of the genomes of wealthy individuals from the past. And in fact, if inequality is genetically rooted, then the genes associated with it have been increasing in humankind's gene pool to the detriment of other genetic variants, just as adaptive variants to traits like disease resistance and lactose tolerance have been selected for other reasons. These genes could be related to human behavior and cognitive abilities, but some might just be linked to physical traits such as increased height. We can assume that genetic inequality thrives in our genomes, but also that a large part of the current genomic variation of humankind has been contributed by women, likely the largest human group to be systematically subjected to inequality over the past several thousand years.

We are now able to screen for people in specific social and cultural circumstances across time periods, and survey their genetic contributions to subsequent populations. All the little personal decisions and their consequences, one stacked atop another, form a composite picture through the years of the enormous inequalities in ancestry that we've seen throughout this book. Soon we'll start sequencing genomes from the past few hundred years and find genome blocks that are identical to those of modern individuals due to shared ancestry.[8] We'll be able to link the past to the present and understand present-day populations as a complex genealogical—as well as genetic—network anchored in the past.[9] The Icelandic study of Jonathan is just one example of this, but many more will come.[10]

A further genetic lesson from the current pandemic is that once again, wealthy people will tend to survive, and in turn, their genes will tend to be overrepresented in the population after the pandemic. This can happen again and again in the future, such as in crises associated with climate change. Coupled with high rates of assortative mating, the link between inequality and biology will gain strength in the future. If wealthy people tend to mate among them, irrespective of their population background, inequality would likely erase genetic differences among the high-status social class. In a globalized world, ethnic identity will be left to the lower classes of society.

A few years ago, at the peak of globalization, some intellectuals defended the view that genetic differences between populations would decrease and humanity would move toward a genetic—as well as physical—uniformity. The proponents of this vision went so far as to predict the future appearance of humankind, saying that people would have beige or brown skin (somewhere between the dark tones of Africans and the fair complexions of Europeans and Asians).[11] Of course this vision disregarded the strong evidence for assortative mating that we've seen to exist throughout this book. But to assess the impact of globalization we must also look back and consider what happened in other historical periods where large populations were connected under the same social system (even if not at the geographic scale of today). Probably the Roman Empire is the most obvious example of a large area under a single political command, and we are now, two thousand years later, starting to explore genomes from that period across three continents.

In a study of 127 ancient genomes from the Lazio region, researchers described how by the time of Rome's foundation, the ancestry of its inhabitants was similar to that of people who settled around the Mediterranean region after the population movements of the Bronze and Iron Ages.[12] In contrast, at the height of the empire, ancestry shifted toward Near Eastern populations and became much more diverse too. Eastern Mediterranean cities such as Ephesus, Alexandria, and Antioch were, after the capital, the most densely populated of the empire, and Rome received immigrants from the eastern provinces that were also the richest regions. But afterward, during the period of late antiquity and the subsequent Barbarian invasions, ancestry shifted again toward central and northern Europe; finally, with the dissolution of the Western Roman Empire, the genetic composition became quite similar to what is seen today in the region. Surprising as it may seem for ancestry to experience such a substantial shift in a few centuries, we have to consider that by definition, large cities are more susceptible to famines, pandemics, wars, and depopulation crises than the less populated but far more stable countryside. Additional studies on Roman settlements in other provinces attest as well to this enormous cosmopolitanism, which later vanished. For instance, a possible Roman gladiator excavated in York (Great Britain) had clear genetic affinities to modern Jordanian people. The genetic analysis of several dozen individuals from the Roman military city of Viminacium, in present-day Serbia, also shows an enormous level of genetic heterogeneity in that fairly remote spot, including an East African male.[13] The overall genetic affinities of this outpost located at the Danubian border again shifted toward the Near East, and again this diversity collapsed with the migrations into the territory of Goths and Slavs, whose arrival shaped the genomes of the locals into something quite similar to those of modern Serbs. So globalization comes and goes. While it affects genetics, it doesn't appear to leave the expected long-term marks in genomic uniformization. The same can occur in our own populations, which may no longer be so global after the coronavirus pandemic. What can be said confidently is that assortative mating is still operating in our societies and that inequality might even increase it.

There is the possibility too, more than ever before, that inequality will again become biological because wealthy people might opt to edit their genomes in specific genes, such as to acquire resistance to future pandemics or improve certain cognitive abilities. Genetic differences within societies could increase as the gap in income widens.

Intellectuals exploring dystopian futures have already pointed to the potential link between social class and biology. In *Brave New World* (the title is taken from a line from Shakespeare's *The Tempest*), the renowned book about a dystopian society by Aldous Huxley, social stratification is associated with biological changes. (The idea may come as no surprise when you realize that the author was the grandson of English naturalist Charles Darwin's friend Thomas Henry Huxley and that his brother, Julian Huxley, was a renowned biologist.) Published in 1932, the novel is set in London in the year 2540, when humans are grown in artificial uteri, and assigned to predetermined social classes according to physical traits such as stature and intelligence. Members of the upper caste, named alpha-plus, are taller and smarter than others, and are assigned to government tasks; below them are the beta, gamma, delta, and epsilon. The lower castes—who are shorter, uglier, and darker—work at the hardest manual tasks. Besides their physical differences, each class is dressed in different colors in order to be easily distinguished.[14] Not only writers have explored the subject; filmmakers have grasped its appeal to audiences. The 1997 film *Gattaca* updates Huxley's sinister notion by presenting a futuristic society in which one's genome determines which jobs one can aspire to, ultimately making the point that the power of the will is more important than one's physical or genetic attributes. The impact of the environment is also made clear in the character of

Jerome Eugene Morrow, whose optimal genetic composition makes him eligible for any job—until a car accident leaves him confined to a wheelchair. Once again we are reminded of the power of nurture over nature.

With the development of gene-editing technologies based on CRISPR, it is now possible to change one's personal genetic composition. Even if this is (or will be) regulated throughout the world's societies, an incident in China demonstrated how easily the germ line can now be modified. In 2018, Chinese researcher He Jiankui edited the genomes of two embryos, in the process making the resulting CCR5 protein nonfunctional.[15] This was an attempt to protect the resulting children against HIV, as it is known that a mutation in the CCR5 gene that disables the CCR5 receptor on the surface of white blood cells confers innate resistance to the virus (the father in the Chinese case was HIV positive).[16] The controversial experiment might have secondarily resulted in impaired brain function and shortened life expectancy, as the CCR5 gene is involved in a range of other functions.

Nevertheless, the He Jiankui affair showed that relatively inexperienced researchers could carry out such a procedure, prompting some scientists—like Josiah Zayner, who called himself a biohacker—to experiment on themselves by injecting CRISPR-modified DNA into their tissue at home.[17] Russian-born millionaire Serge Faguet, who advocates DNA editing for extending one's life, said, "People here [in Silicon Valley] have a technical mindset, so they think of everything as an engineering problem. A lot of people who are not of a technical mindset assume that, 'Hey, people have always been dying,' but I think there's going to be a greater level of awareness [of biohacking] once results start to happen."[18]

It should come as no surprise, then, that wealthy people would choose to modify their own genetic composition prior to adjusting that of their offspring. The most obvious objective would be to extend their life expectancy—in a world of dwindling resources and increasing population growth—and increase their cognitive abilities, including, say, memory enhancement.[19] But it's just a matter of time until they're able to provide their offspring with physical traits associated with social success, such as increased height or lighter pigmentation. While the former is a complex trait that likely depends on some hundreds of genetic variants, the latter is strongly influenced by just a few genes and their modification could likely be accomplished with relatively few gene edits. Furthermore, wealthy people could modify their offspring in aspects associated with socially approved standards of beauty such as blue eyes or red hair, both physical traits that depend essentially on a single genetic modification.[20] We could thus be on the verge of a new CRISPR-based eugenics movement. (Eugenics, a term first coined by Darwin's cousin Francis Galton in 1883, refers to the social practice of "improving" the human species by promoting the selective mating of people with certain desirable traits—or discouraging or simply forbidding the mating of people considered inferior.)[21] In this case, the privileged class might not advocate for the improvement of all society but rather devote its formidable resources and efforts to improving its own family lines.[22] Through the combination of unprecedented wealth and accurate gene editing, the old utopias may evolve in unexpected ways.

The perpetuation of the dynasties of the wealthy, along with similarly acquired genetic modifications, could in the long term create genomic compositions that differ substantially from those of the overall population. This has to some degree already occurred with Europe's royal families, where assortative mating processes through a number of generations have created endogamous genomic compositions that are not at all representative of the population they ruled. In this sense, certain physical traits, such as the famous Habsburg jaw, might have resulted from inbreeding. Researchers have estimated that in family members from the Spanish royal line, on average 9 percent of their genes are identical in both the paternal and maternal chromosomes because they come from the same ancestor.[23] Marrying almost exclusively within the same family in the last generation prior to the last representative of the family, King Charles II of Spain, means that their genomic diversity did not, in many ways, match the diversity found in the country at that time. But avoiding outside marriage and controlling the redistribution of wealth is not exclusive to royal lines. The German French banker Jakob Mayer Rothschild (1792–1868), for instance, married his own niece, and many of his children were wedded to their first cousins: Charlotte de Rothschild to Nathaniel de Rothschild, Mayer Alphonse de Rothschild to Leonora de Rothschild, Salomon James de Rothschild to Adèle von Rothschild, and Edmond Benjamin de Rothschild to Adelheid von Rothschild.[24]

At the level of social classes—including the Indian stratified system— some operating mechanisms such as mating within the same class or even the same family, have decreased genetic diversity in each subgroup to an unexpectedly low level for the population size involved. This in turn can have consequences for health, as many endogamous communities already know. Tay-Sachs disease, for example, has a high incidence among Ashkenazi Jews; one in thirty-one members of the North American Jewish community is a carrier of the mutation causing this fatal genetic disorder in homozygosity.[25] The Amish too suffer a range of rare genetic disorders due to a phenomenon known as a founder effect, which greatly reduces genetic diversity and increases potentially damaging mutations to a high frequency in isolated populations.[26] In Lancaster County, Pennsylvania, for instance, more than fifty thousand Amish can trace their ancestors back to just eighty German Swiss people who settled there in the eighteenth century.[27]

We have seen that wealth can be inherited, even in modern times and Western societies, to a degree that seems to last for generations, if not centuries. In fact we can confidently assume that any wealthy person we come across has inherited their fortune. To the point that wealth exponentially increases in the present day, dynastic lines thriving on this inherited wealth will have a much longer perdurability than similar families in the past. A wealthy family today will be wealthier in the future. And I predict that assortative mating will continue to operate at a higher scale. It is thus not implausible to say that the convergence of wealth and biology can leave a signature that will last well into the future, just as it has in the past.

German poet and dramatist Johann Wolfgang von Goethe had a harrowing dictum: "We all live on the past, and through the past are destroyed."[28] Thanks to the continuing efforts of genomic studies, we'll eventually be able to quantify inequality through history and evaluate its social consequences at different historical moments. Clearly, past inequality lives within us, and we will have to decide how we want to face it. That said, inequality won't be a purely academic discussion about the distant past; if anything, the subject will shape our current social and political debates, and it is likely that only with the integration of all historical, political, cultural, and even behavioral aspects will we be able to properly understand its socioeconomic consequences. The past and future will question us relentlessly, revealing the influence of their invisible yet powerful links, not only in our genomes, but in our minds. Our societies are currently embroiled in dynamics that involve genetics and inequality, and the social responses to these issues are not at all evident or without costs.

Along with such other threats as polarization, disinformation, state violence, and suppression of civil rights, inequality seems to be one of the main obstacles to the survival of democracy, as we understand it, in this century.[29] Now, as the genetic future of inequality stretches before us, besides having more detailed information about its magnitude and consequences than ever before, we have an unprecedented range of social mechanisms to discuss it and propose solutions. This is indeed a challenging time, maybe even a turning point of this century. But it is also a time to act, a time to see beyond the burden of the past.

Acknowledgments

I drafted the main chapters of this book during the severe lockdown of the COVID-19 pandemic that lasted from mid-March until June 2020 in Spain; distressed and stuck at home, it was hardly the best time for any intellectual activity—but we can ask ourselves, When is the right time for such an enterprise anyway? And without the support of my wife, Marta, and my children, Martina and Marc, I would not have been able to finish it.

Different friends and colleagues read this manuscript, or parts of it, and provided useful comments. Any omissions or mistakes as well as interpretations and speculations are just my own responsibility. I am deeply grateful to Agnar Helgason, M. Thomas P. Gilbert, David Reich, Patrick J. Geary, Rob R. Dunn, Arcadi Navarro, Francesc Calafell, Tomàs Marquès-Bonet, Antonio González-Martín, Iñigo Olalde, Vanessa Villalba-Mouco, Ricard Solé, Robert Sala, Xavier Bellés, Miodrag Grbic, and Kristian Kristiansen for critical reading, and also to Thomas Piketty and Samuel S. Bowles, who suggested some additional reading. I am indebted to Daniel Schechter and Cindy Milstein, who edited the whole manuscript, and Jordi Corbera, who created the figures. I am also deeply grateful to the Consejo Superior de Investigaciones Científicas for institutional support, and the Spanish Ministry of Science, Innovation and Universities as well as the Departament d'Universitats i Recerca (Generalitat de Catalunya) for funding my research in paleogenetics, whose results provided me with some of the ideas for this book.

I also would like to thank the MIT Press for its enthusiastic and continuous support of this project, and especially executive editor Robert V. Prior and his assistant editor, Anne-Marie Bono.

Finally, my parents are always in my heart.

Notes

Preface

1. I. Olalde, M. E. Allentoft, F. Sánchez-Quinto, G. Santpere, C. W. K. Chiang, M. DeGiorgio, J. Prado-Martinez, et al., "Derived Immune and Ancestral Pigmentation Alleles in a 7,000-Year-Old Mesolithic European," *Nature* 507 (2014): 225–228; I. Olalde, H. Schroeder, M. Sandoval-Velasco, L. Vinner, I. Lobón, O. Ramirez, S. Civit, et al., "A Common Genetic Origin for Early Farmers from Mediterranean Cardial and Central European LBK Cultures," *Molecular Biology and Evolution* 32 (2015): 3132–3142.

2. D. Reich, *Who We Are and How We Got Here: Ancient DNA and the New Science of the Human Past* (New York: Pantheon Books, 2018).

3. F. Racimo, M. Sikora, M. Vander Linden, H. Schroeder, and C. Lalueza-Fox, "Beyond Broad Strokes: Sociocultural Insights from the Study of Ancient Genomes," *Nature Review Genetics* 21, no. 6 (2020): 355–366.

4. K. Gehred, "Did George Washington's False Teeth Come from His Slaves? A Look at the Evidence, the Responses to That Evidence, and the Limitations of History," Washington Papers, October 19, 2016; J. Van Horn, "George Washington's Dentures: Disability, Deception, and the Republican Body," *Early American Studies* 14 (2016): 2–47.

5. W. Benjamin, *On the Concept of History* (Scotts Valley, CA: CreateSpace, 2016).

Chapter 1

1. S. Meredith, "Total Billionaire Wealth Surges to Record High of 10.2 Trillion Dollars during Coronavirus Crisis, Research Says," CNBC, October 7, 2020.

2. T. Piketty, *Capital in the Twenty-First Century* (Cambridge, MA: Belknap Press of Harvard University Press, 2014).

3. F. Alvaredo, L. Chancel, T. Piketty, E. Saez, and G. Zucman, *World Inequality Report* (Paris: the World Inequality Lab, 2018).

4. Alvaredo, et al., *World Inequality Report.*

5. Piketty, *Capital in the Twenty-First Century*, 190.

6. S. Pinker, *Enlightenment Now: The Case for Reason, Science, Humanism, and Progress* (New York: Viking Books, 2018).

7. I. Goldin and M. Mariathasan, *The Butterfly Defect: How Globalization Creates Systemic Risks, and What to Do about It* (Princeton, NJ: Princeton University Press, 2014).

8. H. Rosling, *Factfulness: Ten Reasons We're Wrong about the World—and Why Things Are Better Than You Think* (London: Hodder and Stoughton, 2018); O. Burkeman, "Is the World Really Better Than Ever?," *Guardian*, July 28, 2017; R. Paulsen, "Why You Shouldn't Listen to Self-Serving Optimists Like Hans Rosling and Steven Pinker," *In These Times*, March 27, 2019.

9. M. Cooper, "The False Promise of Meritocracy," *Atlantic*, December 1, 2015.

10. W. Scheidel, *The Great Leveler: Violence and History of Inequality from the Stone Age to the Twenty-First Century* (Princeton, NJ: Princeton University Press, 2017).

11. O. J. Benedictow, *The Black Death 1346–1353: The Complete History* (Woodbridge, UK: Boydell Press, 2018).

12. G. Alfani and M. Di Tullio, *The Lion's Share: Inequality and the Rise of the Fiscal State in Preindustrial Europe* (Cambridge: Cambridge University Press, 2019).

13. W. Scheidel, *The Great Leveler.*

14. J. Neel, "Is There Hope for the American Dream? What Americans Think about Income Inequality," NPR, January 9, 2020.

15. C. Starmans, M. Sheskin, and P. Bloom, "Why People Prefer Unequal Societies," *Nature Human Behaviour* 1 (2017): 0082.

16. D. Houser and K. McCabe, "Experimental Economics and Experimental Game Theory," in *Neuroeconomics: Decision Making and the Brain*, ed. P. W. Glimcher, C. F. Camerer, E. Fehr, and R. A. Poldrack (Cambridge, MA: Academic Press, 2008), 19–34.

17. M. Schäfer, D. B. M. Haun, and M. Tomasello, "Fair Is Not Fair Everywhere," *Psychological Science* 26 (2015): 1252–1260; J. H. Barkow, L. Cosmides, and J. Tooby, eds., *The Adapted Mind: Evolutionary Psychology and the Generation of Culture* (Oxford: Oxford University Press, 1992).

18. J. Leder and A. Schütz, "Dictator Game," in *Encyclopedia of Personality and Individual Differences*, ed. V. Zeigler-Hill and T. K. Shackelford (Cham, Switzerland: Springer, 2018).

19. S. F. Brosnan and B. M. de Waal, "Monkeys Reject Unequal Pay," *Nature* 425 (2003): 297–299.

20. S. Dobove, N. Baumard, and J.-B. André, "On the Evolutionary Origins of Equity," *PLoS ONE* 12 (2017): e0173636.

21. H. R. Hermann, *Dominance and Aggression in Humans and Other Animals: The Great Game of Life* (Cambridge, MA: Academic Press, 2017).

22. R. W. Wrangham, "Two Types of Aggression in Human Evolution," *Proceedings of the National Academy of Sciences USA* 115 (2018): 245–253.

23. S. Bowles, "Did Warfare among Ancestral Hunter-Gatherers Affect the Evolution of Human Social Behaviors?," *Science* 324 (2009): 1293–1298; M. Lipson, I. Ribot, S. Mallick, N. Rohland, I. Olalde, N. Adamski, N. Broomandkhoshbacht, et al., "Ancient West African Foragers in the Context of African Population History," *Nature* 577 (2020): 665–670; W. Ke, et al., "Ancient Genomes Reveal Complex Patterns of Population Movement, Interaction, and Replacement in sub-Saharan Africa," *Science Advances* 6, no. 24 (2020): eaaz0183.

24. M. Daly, "Partitioning Aggression," *Proceedings of the National Academy of Sciences USA* 115 (2018): 633–634.

25. R. E. Green, "A Draft Sequence of the Neanderthal Genome," *Science* 328 (2010): 710–722; D. Reich, R. E. Green, M. Kircher, J. Krause, N. Patterson, E. Y. Durand, B. Viola, et al., "Genetic History of an Archaic Hominin Group from Denisova Cave in Siberia," *Nature* 468 (2010): 1053–1060.

26. F. L. Mendez, G. D. Poznik, S. Castellano, and C. D. Bustamante, "The Divergence of Neandertal and Modern Human Y Chromosomes," *American Journal of Human Genetics* 98 (2016): 728–734; M. Petr, M. Hajdinjak, Q. Fu, E. Essex, H. Rougier, I. Crevecoeur, P. Semal, et al., "The Evolutionary History of Neanderthal and Denisovan Y Chromosomes," *Science* 369 (2020): 1653–1656.

27. C. Lalueza-Fox, A. Rosas, A. Estalrrich, E. Gigli, P. F. Campos, A. García-Tabernero, S. García-Vargas, et al., "Genetic Evidence for Patrilocal Mating behavior among Neandertal Groups," *Proceedings of the National Academy of Sciences USA* 108 (2011): 250–253.

28. S. Pinker, *The Better Angels of Our Nature: Why Violence Has Declined* (New York: Viking Books, 2011).

29. S. Bowles, "Did Warfare among Ancestral Hunter-Gatherers Affect the Evolution of Social Behaviors?," *Science* 324 (2009): 1293–1298.

30. D. P. Fry and P. Söderberg, "Lethal Aggression in Mobile Forager Bands and Implications for the Origins of War," *Science* 341 (2013): 270–273.

31. P. Cirillo and N. N. Taleb, "The Decline of Violent Conflicts: What Do The Data Really Say?," in *Nobel Foundation Symposium 161: The Causes of Peace* (Stockholm: Nobel Foundation, 2016).

32. N. C. Kim and M. Kissel, *Emergent Warfare in Our Evolutionary Past* (New York: Routledge, 2018); N. Sala, J. L. Arsuaga, A. Pantoja-Pérez, A. Pablos, I. Martínez, R. M. Quam, A. Gómez-Olivencia, et al., "Lethal Interpersonal Violence in the Middle Pleistocene," *PLoS ONE* 10 (2015): e0126589.

33. J.-K. Choi and S. Bowles, "The Coevolution of Parochial Altruism and War," *Science* 318 (2007): 636–640.

34. M. Borgerhoff and B. A. Beheim, "Understanding the Nature of Wealth and Its Effects on Human Fitness," *Philosophical Transactions of the Royal Society B* 366 (2011): 344–356.

35. S. Bowles, E. A. Smith, and M. Borgerhoff, "The Emergence and Persistence of Inequality in Premodern Societies: Introduction to the Special Section," *Current Anthropology* 51 (2010): 7–17.

36. M. Dyble, G. D. Salali, N. Chaudhary, A. Page, D. Smith, J. Thompson, L. Vinicius, et al., "Sex Equality Can Explain the Unique Social Structure of Hunter-Gatherer Bands," *Science* 348 (2015): 796–798.

37. B. Hayden, *The Pithouses of Keatley Creek: Complex Hunter-Gatherers of the Northwest Plateau* (Burnaby, BC: SFU Archaeology Press, 2005); Bowles, Smith, and Borgerhoff, "The Emergence and Persistence of Inequality in Premodern Societies."

38. E. Trinkaus and A. P. Buzhilova, "Diversity and Differential Disposal of the Dead at Sunghir," *Antiquity* 92 (2018): 7–21.

39. L. Geggel, "Ochre: The World's First Red Paint," *LiveScience*, November 20, 2018.

40. A. F. Dixson and B. J. Dixson, "Venus Figurines of the European Paleolithic: Symbols of Fertility or Attractiveness?," *Journal of Anthropology* 2011 (2012): 569120.

41. K. R. Vandewettering, "Upper Paleolithic Venus Figurines and Interpretations of Prehistoric Gender Representations," *PURE Insights* 4 (2015): article 7.

42. B. Klima, "A Triple Burial from the Upper Paleolithic of Dolní Věstonice, Czechoslovakia," *Journal of Human Evolution* 16, nos. 7–8 (November–December 1987): 831–835.

43. V. Formicola, A. Pontrandolfi, and J. Svoboda, "The Upper Paleolithic Triple Burial of Dolní Věstonice: Pathology and Funerary Behavior," *American Journal of Physical Anthropology* 115, no. 4 (2001): 372–379.

44. Q. Fu, C. Posth, M. Hajdinjak, M. Petr, S. Mallick, D. Fernandes, A. Furtwängler, et al., "The Genetic History of Ice Age Europe," *Nature* 534 (2016): 200–205.

45. M. Sikora, A. Seguin-Orlando, V. C. Sousa, A. Albrechtsen, T. Korneliussen, A. Ko, S. Rasmussen, et al., "Ancient Genomes Show Social and Reproductive Behavior of Early Upper Paleolithic Foragers," *Science* 358 (2017): 659–662.

46. G. Clark, "Sweden: Mobility Achieved?," in *The Son Also Rises: Surnames and the History of Social Mobility* (Princeton, NJ: Princeton University Press, 2014), 19–44.

47. G. Clark, "Medieval England: Mobility in the Feudal Age," in *The Son Also Rises: Surnames and the History of Social Mobility* (Princeton, NJ: Princeton University Press, 2014), 70–87.

48. F. Torche and A. Corvalan, "Estimating Intergenerational Mobility with Grouped Data: A Critique of Clark's the Son Also Rises," *Sociological Methods and Research* 47 (2018): 787–811.

49. Organization for Economic Cooperation and Development, *Education at a Glance 2011: OECD Indicators* (Paris: OECD Publishing, 2011).

50. F. Calafell, personal communication. Data from INE, 2010.

51. G. Clark, "The Condition of the Working Class in England, 1209–2004," *Journal of Political Economy* 113 (2005): 1307–1340.

52. A. Jagadeesan, E. D. Gunnarsdóttir, S. S. Ebenesersdóttir, V. B. Guðmundsdóttir, E. L. Thordardóttir, M. S. Einarsdóttir, H. Jónsson, J.-M. Dugoujon, et al., "Reconstructing an African Haploid Genome from the 18th Century," *Nature Genetics* 50 (2018): 199–205.

53. G. Pálsson, *The Man Who Stole Himself: The Slave Odyssey of Hans Jonathan* (Chicago: University of Chicago Press, 2016).

Chapter 2

1. D. Reich, *Who We Are and How We Got Here: Ancient DNA and the New Science of the Human Past* (New York: Pantheon Books, 2018).

2. S. Marciniak and G. H. Perry, "Harnessing Ancient Genomes to Study the History of Human Adaptation," *Nature Reviews Genetics* 18 (2017): 659–674. Estimates updated with 2018–2019 data.

3. J.-J. Rousseau, *A Discourse on Inequality* (Indianapolis, Indiana: Hackett Publishing, 1992), 44.

4. Rousseau, *A Discourse on Inequality*, 44.

5. F. Pessoa, *The Book of Disquiet* (London: Quartet Books, 1991).

6. L. Damrosch, *Jean-Jacques Rousseau: Restless Genius* (New York: Houghton Mifflin Harcourt, 2005), 488.

7. R. Douglass, *Rousseau and Hobbes: Nature, Free Will, and the Passions* (Oxford: Oxford University Press, 2015).

8. R. Wrangham, *The Goodness Paradox: The Strange Relationship between Virtue and Violence in Human Evolution* (New York: Pantheon, 2019).

9. V. G. Childe, "Changing Methods and Aims in Prehistory: Presidential Address for 1935," *Proceedings of the Prehistoric Society* 1 (2014): 1–15.

10. J. Diamond, *Guns, Germs, and Steel: The Fates of Human Societies* (New York: W. W. Norton and Company, 1997).

11. S. Svizzero, "Persistent Controversies about the Neolithic Revolution," *Journal of Historical Archaeology and Anthropologal Sciences* 1 (2017): 53–61.

12. A. Mummert, E. Esche, J. Robinson, and G. J. Armelagos, "Stature and Robusticity during the Agricultural Transition: Evidence from the Bioarchaeological Record," *Economics and Human Biology* 9 (2011): 284–301.

13. J. C. Scott, *Against the Grain: A Deep History of the Earliest States* (New Haven: Yale University Press, 2017).

14. M. Hermanussen, "Stature of Early Europeans," *Hormones* 2 (2003): 175–178.

15. I. Olalde, M. E. Allentoft, F. Sánchez-Quinto, G. Santpere, C. W. K. Chiang, M. DeGiorgio, J. Prado-Martinez, et al., "Derived Immune and Ancestral Pigmentation Alleles in a 7,000-Year-Old Mesolithic European," *Nature* 507 (2014): 225–228.

16. P. Skoglund, H. Malmström, M. Raghavan, J. Storå, P. Hall, E. Willerslev, M. Thomas, et al., "Origins and Genetic Legacy of Neolithic Farmers and Hunter-Gatherers in Europe," *Science* 336 (2012): 466–469.

17. G. Brandt, W. Haak, C. J. Alder, C. Roth, A. Szécsényi-Nagy, S. Karimnia, S. Möller-Rieker, et al., "Ancient DNA Reveals Key Stages in the Formation of Central European Mitochondrial Genetic Diversity," *Science* 342 (2013): 257–261.

18. S. Brace, Y. Diekmann, T. J. Booth, L. van Dorp, Z. Faltyskova, N. Rohland, S. Mallick, et al., "Ancient Genomes Indicate Population Replacement in Early Neolithic Britain," *Nature Ecology and Evolution* 3 (2019): 765–771.

19. I. Mathieson, S. Alpaslan-Roodenberg, C. Posth, A. Szécsényi-Nagy, N. Rohland, S. Mallick, I. Olalde, et al., "The Genomic History of Southeastern Europe," *Nature* 555 (2018): 197–203.

20. C. Gamba, E. R. Jones, M. D. Teasdale, R. L. McLaughlin, G. Gonzalez-Fortes, V. Mattiangeli, L. Domboróczki, et al., "Genome Flux and Stasis in a Five Millennium Transect of European Prehistory," *Nature Communications* 5 (2014): 5257.

21. Mathieson, et al., "The Genomic History of Southeastern Europe."

22. I. Olalde, H. Schroeder, M. Sandoval-Velasco, L. Vinner, I. Lobón, O. Ramirez, S. Civit, et al., "A Common Genetic Origin for Early Farmers from Mediterranean Cardial and Central European LBK Cultures," *Molecular Biology and Evolution* 32 (2015): 3132–3142.

23. C. Valdiosera, T. Günther, J. C. Vera-Rodríguez, I. Ureña, E. Iriarte, R. Rodríguez-Varela, L. G. Simões, et al., "Four Millennia of Iberian Biomolecular Prehistory Illustrate the Impact of Prehistoric Migrations at the Far End of Eurasia," *Proceedings of the National Academy of Sciences USA* 115 (2018): 3428–3433.

24. M. Lipson, A. Szécsényi-Nagy, S. Mallick, A. Pósa, B. Stégmár, V. Keerl, N. Rohland, et al., "Parallel Palaeogenomic Transects Reveal Complex Genetic History of Early European Farmers," *Nature* 551 (2017): 368–372.

25. Mathieson, et al., "The Genomic History of Southeastern Europe."

26. Brace, et al., "Ancient Genomes Indicate Population Replacement in Early Neolithic Britain."

27. D. M. Fernandes, D. Strapagiel, P. Borówka, B. Marciniak, E. Żądzińska, K. Sirak, V. Siska, et al., "A Genomic Neolithic Time Transect of Hunter-Farmer Admixture in Central Poland," *Scientific Reports* 8 (2018): 14879.

28. Brace, "Ancient Genomes Indicate Population Replacement in Early Neolithic Britain."

29. I. Lazaridis, N. Patterson, A. Mittnik, G. Renaud, S. Mallick, K. Kirsanow, P. H. Sudmant, et al., "Ancient Human Genomes Suggest Three Ancestral Populations for Present-Day Europeans," *Nature* 513 (2014): 409–413.

30. S. Elliott, "The Walls That Did Not Come Tumbling Down: Are the Early Neolithic Walls of Jericho the First Evidence of Warfare?," *RUSI Journal* 157 (2012): 72–79.

31. C. Meyer, C. Lohr, D. Gronenborn, and K. W. Alt, "The Massacre Mass Grave of Schöneck-Kilianstädten Reveals New Insights into Collective Violence in Early Neolithic Central Europe," *Proceedings of the National Academy of Sciences USA* 112 (2015): 11217–11222.

32. C. Meyer, C. Knipper, N. Nicklisch, A. Münster, O. Kürbis, V. Dresely, H. Meller, and K. W. Alt, "Early Neolithic Executions Indicated by Clustered Cranial Trauma in the Mass Grave of Halberstadt," *Nature Communications* 9 (2018): 2472.

33. M. Mirazón Lahr, F. Rivera, R. K. Power, A. Mounier, B. Copsey, F. Crivellaro, J. E. Edung, et al., "Inter-Group Violence among Early Holocene Hunter-Gatherers of West Turkana, Kenya," *Nature* 529 (2016): 394–398.

34. S. Gummesson, F. Hallgren, and A. Kjellström, "Keep Your Head High: Skulls on Stakes and Cranial Trauma in Mesolithic Sweden," *Antiquity* 92 (2018): 74–90.

35. Lazaridis, et al., "Ancient Human Genomes Suggest Three Ancestral Populations for Present-Day Europeans."

36. W. Haak, I. Lazaridis, N. Patterson, N. Rohland, S. Mallick, B. Llamas, G. Brandt, et al., "Massive Migration from the Steppe was a Source for Indo-European Languages in Europe," *Nature* 522 (2015): 207–211; M. E. Allentoft, M. Sikora, K.-G. Sjögren, S. Rasmussen, M. Rasmussen, J. Stenderup, P. B. Damgaard, et al., "Population Genomics of Bronze Age Eurasia," *Nature* 522 (2015): 167–172.

37. N. Gogol, *Taras Bulba and Other Tales*, Project Gutenberg, e-book #1197, 2017, www.gutenberg.org/1197/1197-h/1197-h.htm.

38. N. Rascovan, K.-G. Sjögren, K. Kristiansen, R. Nielsen, E. Willerslev, C. Desnues, S. Rasmussen, et al., "Emergence and Spread of Basal Lineages of *Yersinia pestis* during the Neolithic Decline," *Cell* 176 (2019): 295–305.

39. D. W. Anthony, *The Horse, the Wheel, and Language* (Princeton, NJ; Princeton University Press, 2007).

40. P. de Barros Damgaard, R. Martiniano, J. Kamm, J. V. Moreno-Mayar, G. Kroonen, M. Peyrot, G. Barjamovic, et al., "The First Horse Herders and the Impact of Early Bronze Age Steppe Expansions into Asia," *Science* 360 (2018): eaar7711.

41. Anthony, *The Horse, the Wheel, and Language*.

42. K. Kristiansen, "The Bronze Age Expansion of Indo-European Languages: An Archaeological Model," in *Becoming European: The Transformation of Third Millennium Northern and Western Europe*, ed. C. Prescott and H. Glorstad (Oxford: Oxbow Books, 2011), 165–181.

43. K. Kristiansen, M. E. Allentoft, K. M Frei, R. Iversen, N. N. Johannsen, G. Kroonen, Ł. Pospieszny, T. D. Price, et al., "Re-Theorising Mobility and the Formation of Culture and Language among the Corded Ware Culture in Europe," *Antiquity* 91 (2017): 334–347.

44. Lazaridis, et al., "Ancient Human Genomes Suggest Three Ancestral Populations for Present-Day Europeans."

45. Y. Itan, A. Powell, M. A. Beaumont, J. Burger, and M. G. Thomas, "The Origins of Lactase Persistence in Europe," *PLoS Computational Biology* 5 (2009): e1000491.

46. I. Mathieson, I. Lazaridis, N. Rohland, S. Mallick, N. Patterson, S. A. Roodenberg, et al., "Genome-Wide Patterns of Selection in 230 Ancient Eurasians," *Nature* 528 (2015): 499–503.

47. L. Yengo, J. Sidorenko, K. E. Kemper, Z. Zheng, A. R. Wood, M. N. Weedon, T. M. Frayling, et al. "Meta-analysis of Genome-Wide Association Studies for Height and Body Mass Index in ~700000 Individuals of European Ancestry," *Human Molecular Genetics* 15 (2018): 3641–3649.

48. H. Schroeder, A. Margaryan, M. Szmyt, B. Theulot, P. Włodarczak, S. Rasmussen, S. Gopalakrishnan, et al., "Unraveling Ancestry, Kinship, and Violence in a Late Neolithic Mass Grave," *Proceedings of the National Academy of Sciences USA* 116 (2019): 10705–10710.

49. S. Rasmussen, M. E. Allentoft, K. Nielsen, L. Orlando, M. Sikora, K.-G. Sjögren, A. G. Pedersen, et al., "Early Divergent Strains of *Yersinia pestis* in Eurasia 5,000 Years Ago," *Cell* 163 (2015): P571–P582.

50. Rascovan, et al., "Emergence and Spread of Basal Lineages of *Yersinia pestis* during the Neolithic Decline."

51. J. Czebreszuk, ed., *Similar but Different* (Leiden: Sidestone Press, 2004); H. Fokkens and F. Nicolis, eds., *Background to Beakers; Inquiries into Regional Cultural Backgrounds of the Bell Beaker Complex* (Leiden: Sidestone Press, 2012).

52. I. Olalde, S. Brace, M. E. Allentoft, I. Armit, K. Kristiansen, T. Booth, N. Rohland, et al., "The Beaker Phenomenon and the Genomic Transformation of Northwest Europe," *Nature* 555 (2018): 190–196.

53. N. Patterson, M. Isakov, T. Booth, L. Büster, C.-E. Fischer, I. Olalde, H. Ringbauer, et al., "Large Scale Migration into Southern Britain at the End of the Bronze Age," *Nature* (in press).

54. I. Olalde, S. Mallick, N. Patterson, N. Rohland, V. Villalba-Mouco, M. Silva, K. Dulias, et al., "The Genomic History of the Iberian Peninsula over the Last 8,000 Years," *Science* 363 (2019): 1230–1234.

55. C. Renfrew, *Archaeology and Language: The Puzzle of Indo-European Origins* (London: Jonathan Cape, 1987).

56. Anthony, *The Horse, the Wheel, and Language.*

57. Olalde, et al., "The Genomic History of the Iberian Peninsula over the Last 8,000 Years."

58. C. Ning, C.-C. Wang, S. Gao, Y. Yang, X. Zhang, X. Wu, F. Zhang, et al., "Ancient Genomes Reveal Yamnaya-Related Ancestry and a Potential Source of Indo-European Speakers in Iron Age Tianshan," *Current Biology* 29 (2019): 2526–2532.e4.

59. de Barros Damgaard, et al., "The First Horse Herders and the Impact of Early Bronze Age Steppe Expansions into Asia."

60. Olalde, et al., "The Genomic History of the Iberian Peninsula over the Last 8,000 Years."

61. Reich, *Who We Are and How We Got Here.*

Chapter 3

1. Quoted in C. Desroches-Noblecourt, *Tutankhamen: Life and Death of a Pharaoh* (London: Penguin Books, 1989).

2. C. W. Ceram, *Gods, Graves and Scholars: The Story of Archaeology* (New York: Vintage, 1986).

3. T. A. Kohler, M. E. Smith, A. Bogaard, G. M. Feinman, C. E. Peterson, A. Betzenhauser, M. Pailes, et al., "Greater Post-Neolithic Wealth Disparities in Eurasia Than in North America and Mesoamerica," *Nature* 551 (2017): 619–622.

4. L. Grosman and N. D. Munro, "A Natufian Ritual Event," *Current Anthropology* 57 (2016): 311–331.

5. Kohler, et al., "Greater Post-Neolithic Wealth Disparities in Eurasia Than in North America and Mesoamerica."

6. Kohler, et al., "Greater Post-Neolithic Wealth Disparities in Eurasia Than in North America and Mesoamerica."

7. A. Bogaard, M. Fochesato, and S. Bowles, "The Farming-Inequality Nexus: New Insights from Ancient Western Eurasia," *Antiquity* 93 (2019): 1129–1143.

8. T. Higham, J. Chapman, V. Slavchev, B. Gaydarska, N. Honch, Y. Yordanov, and B. Dimitrova, "New Perspectives on the Varna Cemetery (Bulgaria)—AMS Dates and Social Implications," *Antiquity* 81 (2007): 640–654.

9. A. Fitzpatrick, "The Amesbury Archer: A Well-Furnished Early Bronze Age Burial in Southern England," *Antiquity* 76 (2015): 629–630.

10. H. Vandkilde, "Bronzization: The Bronze Age as Pre-Modern Globalization," *Prähistorische Zeitschrift* 91 (2016): 103–223.

11. V. Lull, R. Micó, C. R. Herrada, R. Risch, E. Celdrán, M. I. Fregeiro, C. Oliart, and C. Velasco, "La Almoloya (Pliego-Muía, Murcia): Palacios y élites gobernantes en la Edad del Bronce," in *El legado de Mula en la historia*, ed. J. A. Zapara Parra (Mula, Spain: Ayuntamiento de Mula, 2016), 41–59.

12. V. Lull, R. Micó, C. Rihuete-Herrada, and R. Risch, "The La Bastida Fortification: New Light and New Questions on Early Bronze Age Societies in the Western Mediterranean," *Antiquity* 88 (2014): 395–401.

13. Kohler, et al., "Greater Post-Neolithic Wealth Disparities in Eurasia Than in North America and Mesoamerica."

14. G. Uslu, O. F. Serifoglu, and R. Van Beek, *Troy: City, Homer and Turkey* (Zwolle, Netherlands: Waanders BV, 2013).

15. A. Stevens, "Death and the City: The Cemeteries of Amarna in Their Urban Context," *Cambridge Archaeological Journal* 28 (2018): 103–126.

16. C. Knipper, M. Fragata, N. Nicklisch, A. Siebert, A. Szécsényi-Nagy, V. Hubensack, C. Metzner-Nebelsick, et al., "A Distinct Section of the Early Bronze Age Society? Stable Isotope Investigations of Burials in Settlement Pits and Multiple Inhumations of the Únêtice Culture in Central Germany," *American Journal of Physical Anthropology* 159 (2015): 496–516.

17. F. Sánchez-Quinto, H. Malmström, M. Fraser, L. Girdland-Flink, E. M. Svensson, L. G. Simões, R. George, et al., "Megalithic Tombs in Western and Northern Neolithic Europe Were Linked to a Kindred Society," *Proceedings of the National Academy of Sciences USA* 116 (2019): 9469–9474.

18. L. M. Cassidy, R. Ó Maoldúin, T. Kador, A. Lynch, C. Jones, P. C. Woodman, E. Murphy, et al., "A Dynastic Elite in Monumental Neolithic Society," *Nature* 582 (2020): 384–388.

19. A. Mittnik, K. Massey, C. Knipper, F. Wittenborn, R. Friedrich, S. Pfrengle, M. Burri, et al., "Kinship-Based Social Inequality in Bronze Age Europe," *Science* 366 (2019): 731–734.

20. Mittnik, et al., "Kinship-Based Social Inequality in Bronze Age Europe."

21. K.-G. Sjögren, I. Olalde, S. Carver, M. E. Allentoft, T. Knowles, G. Kroonen, A. W. G. Pike, et al., "Kinship and Social Organization in Copper Age Europe: A Cross-Disciplinary Analysis of Archaeology, DNA, Isotopes, and Anthropology from Two Bell Beaker Cemeteries," *PLoS ONE* 15 (2020): e0241278.

22. A. Furtwängler, A. B. Rohrlach, T. C. Lamnidis, L. Papac, G. U. Neumann, I. Siebke, E. Reiter, et al., "Ancient Genomes Reveal Social and Genetic Structure of Late Neolithic Switzerland," *Nature Communications* 11 (2020): 1915.

23. J. Burger, V. Link, J. Blöcher, A. Schulz, C. Sell, Z. Pochon, Y. Diekmann, et al., "Low Prevalence of Lactase Persistence in Bronze Age Europe Indicates Ongoing Strong Selection over the Last 3,000 Years," *Current Biology* 30 (2020): P4307–4315.

24. V. Villalba-Mouco, C. Oliart, C. Rihuete-Herrada, A. Childebayeva, A. B. Rohrlach, M. I. Fregeiro, E. Celdrán, et al., "Genomic Transformation and Social Organization during the Copper Age–Bronze Transition in Southern Iberia," *Science Advances* (in press).

25. V. Villalba-Mouco, personal communication.

26. A. P. Fitzpatrick, *The Amesbury Archer and the Boscombe Bowmen: Bell Beaker Burials at Boscombe Down, Amesbury, Wiltshire* (Salisbury, UK: Trust for Wessex Archaeology Ltd., 2011).

27. Sjögren, et al., "Kinship and Social Organization in Copper Age Europe."

28. I. Olalde, S. Brace, M. E. Allentoft, I. Armit, K. Kristiansen, T. Booth, N. Rohland, et al., "The Beaker Phenomenon and the Genomic Transformation of Northwest Europe," *Nature* 555 (2018): 190–196.

29. D. Reich, personal communication.

30. P. Barceló, *Hannibal: Stratege und Staatsmann* (Ditzingen, Germany: Reclam Philipp Jun, 2012).

31. C. E. G. Amorim, S. Vai, C. Posth, A. Modi, I. Koncz, S. Hakenbeck, M. C. La Rocca, et al., "Understanding 6th-Century Barbarian Social Organization and Migration through Paleogenomics," *Nature Communications* 9 (2018): 3547.

32. P. J. Geary, *The Myth of Nations: The Medieval Origins of Europe* (Princeton, NJ: Princeton University Press, 2003).

33. P. Heather, *Empires and Barbarians: The Fall of Rome and the Birth of Europe* (Oxford: Oxford University Press, 2009).

34. S. Eisenmann, E. Bánffy, P. van Dommelen, K. P. Hofmann, J. Maran, I. Lazaridis, A. Mittnik, et al., "Reconciling Material Cultures in Archaeology with Genetic Data: The Nomenclature of Clusters Emerging from Archaeogenomic Analysis," *Scientific Reports* 8 (2018): 13003.

35. M. L. Antonio, Z. Gao, H. M. Moots, M. Lucci, F. Candilio, S. Sawyer, V. Oberreiter, et al., "Ancient Rome: A Genetic Crossroads of Europe and the Mediterranean," *Science* 366 (2019): 708–714.

36. A. Helgason, E. Hickey, S. Goodacre, V. Bosnes, K. Stefánsson, R. Ward, and B. Sykes, "mtDNA and the Islands of the North Atlantic: Estimating the Proportions of Norse and Gaelic Ancestry," *American Journal of Human Genetics* 68 (2001): 723–737.

37. S. S. Ebenesersdóttir, M. Sandoval-Velasco, E. D. Gunnarsdóttir, A. Jagadeesan, V. B. Guðmundsdóttir, E. L. Thordardóttir, M. S. Einarsdóttir, et al., "Ancient Genomes from Iceland Reveal the Making of a Human Population," *Science* 360 (2018): 1028–1032.

38. A. Margaryan, D. J. Lawson, M. Sikora, F. Racimo, S. Rasmussen, I. Moltke, L. M. Cassidy, et al., "Population Genomics of the Viking World," *Nature* 585 (2020): 390–396.

39. S. L. Dawdy, "Archaeology of Modern American Death: Grave Goods and Blithe Mementoes," in *The Oxford Handbook of the Archaeology of the Contemporary World*, ed. P. Graves-Brown, R. Harrison, and A. Piccini (Oxford: Oxford University Press, 2013), 451–466.

40. Quoted in J. Henley, "'Equality Won't Happen by Itself': How Iceland Got Tough on Gender Pay Gap," *Guardian*, February 20, 2018.

Chapter 4

1. Boethius, *The Consolation of Philosophy*, Project Gutenberg, e-book #14328, 2004, www.gutenberg.org/ebooks/14328.

2. T. Piketty, *Capital and Ideology* (Cambridge, MA: Harvard University Press, 2019).

3. G. Clark, *The Son Also Rises: Surnames and the History of Social Mobility* (Princeton, NJ: Princeton University Press, 2014).

4. H. Sherwood, "The Kibbutz: 100 Years Old and Facing an Uncertain Future," *Guardian*, August 13, 2010.

5. M. Shermer, "The Problem with Utopias," *Week*, April 1, 2018.

6. Plato, *The Republic*, trans. G. M. A. Grube (Indianapolis: Hackett Classics, 1992).

7. J. Annas, "Plato's 'Republic' and Feminism," *Phylosophy* 51 (1976): 307–321.

8. E. Brown, "Plato's Ethics and Politics in the Republic," *Stanford Encyclopedia of Phylosophy*, September 12, 2017.

9. R. Miller, "The Utopia for All—with Exceptions: Gender Roles in Thomas More's Utopia and Early Modern England," *Armstrong Undergraduate Journal of History* 9 (2019): article 10.

10. T. More, *Utopia* (New Haven: Yale University Press, 2001).

11. T. Campanella, *The City of the Sun* (Scotts Valley, CA: CreateSpace, 2015).

12. J. V. Andreae, *Christianopolis: An Ideal of the 17th Century*, trans. F. E. Held (New York: Cosimo Classics, 2007).

13. F. Bacon, *The New Atlantis: A Utopian Novel* (Scotts Valley, CA: CreateSpace, 2014).

14. C. Fourier, *The Social Destiny of Man: Or, Theory of the Four Movements* (New Delhi: Sagwan Press, 2015).

15. M. Barrero, "El delirio de Nueva Germania," *Zenda*, October 15, 2019.

16. S. Courtois, N. Werth, J.-L. Panné, A. Paczkowski, K. Bartosek, and J.-L. Margolin, *The Black Book of Communism: Crimes, Terror, Repression* (Cambridge, MA: Harvard University Press, 1999).

17. M. Atwood, *The Handmaid's Tale* (New York: Vintage Classics, 2010).

18. Quoted in Arizona State University College of Liberal Arts and Sciences, "What Are the Roots of Gender Inequality? Women's Rights, Race and Reproduction," *Newswise*, June 1, 2012; S. Kitch, *Higher Ground: From Utopianism to Realism in American Feminist Thought and Theory* (Chicago: University of Chicago Press, 2000).

19. J. S. Mill, *On Liberty* (London: Penguin Books, 1975).

20. G. Vlastos, "Does Slavery Exist in Plato's Republic?," *Classical Philology* 63 (1968): 291–295.

21. G. Alfani and R. Frigeni, "Inequality (Un)perceived: The Emergence of a Discourse of Economic Inequality from the Middle Ages to the Age of Revolution," *Journal of European Economic History* 45 (2016): 21–66.

22. G. Alfani and M. Di Tullio, *The Lion's Share: Inequality and the Rise of the Fiscal State in Preindustrial Europe* (Cambridge: Cambridge University Press, 2019).

23. A. Mittnik, K. Massy, C. Knipper, F. Wittenborn, R. Friedrich, S. Pfrengle, M. Burri, et al., "Kinship-Based Social Inequality in Bronze Age Europe," *Science* 366 (2019): 731–734.

24. Piketty, *Capital and Ideology*.

25. L. Dumont, *Homo Hierarchicus: The Caste System and Its Implications* (Chicago: University of Chicago Press, 1981).

26. D.-E. Berg, *Dynamics of Caste and Law: Dalits, Oppression and Constitutional Democracy in India* (Cambridge: Cambridge University Press, 2019).

27. R. Thapar, *Early India: From the Origins to AD 1300* (Berkeley: University of California Press, 2004).

28. Piketty, *Capital and Ideology*.

29. N. B. Dirks, *Castes of Mind: Colonialism and the Making of Modern India* (Princeton, NJ: Princeton University Press, 2011).

30. Berg, *Dynamics of Caste and Law*.

31. G. Arunkumar, D. F. Soria-Hernanz, V. J. Kavitha, V. S. Arun, A. Syama, K. S. Ashokan, K. T. Gandhirajan, et al., "Population Differentiation in Southern India Male Lineages Correlate with Agricultural Expansions Predating the Caste System," *PLoS One* 7 (2012): e50269; S. Sharma, E. Rai, P. Sharma, M. Jena, S. Singh, K. Darvishi, A. K. Bhat, et al., "The Indian Origin of Paternal Haplogroup R1a1* Substantiates the Autochthonous Origin of Brahmins and the Caste System," *Journal of Human Genetics* 54 (2009): 47–55.

32. D. Reich, K. Thangaraj, N. Patterson, A. L. Price, and L. Singh, "Reconstructing Indian Population History," *Nature* 461 (2009): 489–494.

33. P. Moorjani, K. Thangaraj, N. Patterson, M. Lipson, P.-R. Loh, P. Govindaraj, B. Berger, et al., "Genetic Evidence for Recent Population Mixture in India," *American Journal of Human Genetics* 93 (2013): 422–438.

34. V. M. Narasimhan, N. J. Patterson, P. Moorjani, N. Rohland, R. Bernardos, S. Mallick, I. Lazaridis, et al., "The Genomic Formation of South and Central Asia," *Science* 365 (2019): eeat7487.

35. V. Shinde, V. M. Narasimhan, N. Rohland, S. Mallick, M. Mah, M. Lipson, N. Nakatsuka, et al., "An Ancient Harappan Genome Lacks Ancestry from Steppe Pastoralists or Iranian Farmers," *Cell* 179 (2019): P729–P735.

36. "Steppe Migration to India Was between 3500–4000 Years Ago: David Reich," *Economic Times*, October 12, 2019.

37. Sharma, et al., "The Indian Origin of Paternal Haplogroup R1a1* Substantiates the Autochthonous Origin of Brahmins and the Caste System."

38. Narasimhan, et al., "The Genomic Formation of South and Central Asia."

39. S. Bayly, *Caste, Society and Politics in India from the Eighteenth Century to the Modern Age* (Cambridge: Cambridge University Press, 2001).

40. D. M. Figueira, *Aryans, Jews, Brahmins: Theorizing Authority through Myths of Identity* (Albany: State University of New York Press, 2002).

41. N. Nakatsuka, P. Moorjani, N. Rai, B. Sarkar, A. Tandon, N. Patterson, G. S. Bhavani, et al., "The Promise of Disease Gene Discovery in South Asia," *Nature Genetics* 49 (2017): 1403–1407.

42. S. Daniyal, "How Same-Caste Marriages Persisted for Thousands of Years in India—and Are Still Going on Strong," *Scroll.in*, October 5, 2020.

43. H. Ringbauer, M. Steinrücken, L. Fehren-Schmitz, and D. Reich, "Increased Rate of Close-Kin Unions in the Central Andes in the Half Millennium before European Contact," *Current Biology* 30 (2020): R980–R981.

44. E. Arciero, S. A. Dogra, M. Mezzavilla, T. Tsismentzoglou, Q. A. Huang, K. A. Hunt, D. Mason, et al., "Fine-Scale Population Structure and Demographic History of British Pakistanis," bioRxiv, 2020, https://doi.org/10.1101/2020.09.02.279190.

45. GUaRDIAN Consortium, S. Sivasubbu, and V. Scaria, "Genomics of Rare Genetic Diseases—Experiences from India," *Human Genomics* 13 (2019): article 52.

46. H. Thomas, *The Slave Trade: The Story of the Atlantic Slave Trade, 1440–1870* (New York: Pocket Books, 1999).

47. E. D. Domar, "The Causes of Slavery or Serfdom: A Hypothesis," *Journal of Economic History* 30 (1970): 18–32.

48. Piketty, *Capital and Ideology*.

49. Transatlantic Slave Trade Database, Emory University, 2019, www.slavevoyages .org.

50. H. Schroeder, "Genome-Wide Ancestry of 17th-Century Enslaved Africans from the Caribbean," *Proceedings of the National Academy of Sciences USA* 112 (2015): 3669–3673.

51. R. Barquera, T. C. Lamnidis, A. K. Lankapalli, A. Kocher, D. I. Hernández-Zaragoza, E. A. Nelson, A. C. Zamora-Herrera, et al., "Origin and Health Status of First-Generation Africans from Early Colonial Mexico," *Current Biology* 30 (2020): 2078–2091.e11.

52. D. W. Hill, S. P. Hagenaars, R. E. Marioni, S. E. Harris, D. C. M. Liewald, G. Davies, A. Okbay, et al., "Molecular Genetic Contributions to Social Deprivation and Household Income in UK Biobank," *Current Biology* 26 (2016): 3083–3089.

53. R. E. Marioni, G. Davies, C. Hayward, D. C. M. Liewald, S. M. Kerr, A. Campbell, M. Luciano, et al., "Molecular Genetic Contributions to Socioeconomic Status and Intelligence," *Intelligence* 44 (2014): 26–32.

54. A. Kong, M. L. Frigge, G. Thorleifsson, H. Stefansson, A. I. Young, F. Zink, G. A. Jonsdottir, et al., "Selection against Variants in the Genome Associated with Educational Attainment," *Proceedings of the National Academy of Sciences USA* 114 (2017): E727–E732.

Chapter 5

1. P. J. Gagne, *King's Daughters and Founding Mothers: The Filles du Roi, 1663–1673* (Waterloo, QC: Quintin Publications, 2001).

2. C. S. Larsen, "Equality for the Sexes in Human Evolution? Early Hominid Sexual Dimorphism and Implications for Mating Systems and Social Behavior," *Proceedings of the National Academy of Sciences USA* 100 (2003): 9103–9104.

3. L. Cronk, "Wealth, Status, and Reprodutive Success among the Mukogodo of Kenya," *American Anthropologist* 93 (1991): 345–360; M. Borgerhoff Mulder, I. Fazzio, W. Irons, R. L. McElreath, S. Bowles, A. Bell, T. Hertz, and L. Hazzah, "Pastoralism and Wealth Inequality: Revisiting an Old Question," *Current Anthropology* 51 (2010): 35–48.

4. N. E. Johnson and K. T. Zhang, "Matriarchy, Polyandry, and Fertility among the Mosuos in China," *Journal of Biosocial Science* 23 (1991): 499–505.

5. K. H. Miga, S. Koren, A. Rhie, M. R. Vollger, A. Gershman, A. Bzikadze, S. Brooks, et al., "Telomere-to-Telomere Assembly of a Complete Human X Chromosome," *Nature* 585 (2020): 79–84.

6. A. Goldberg, R. Günther, N. A. Rosenberg, and M. Jakobsson, "Ancient X Chromosomes Reveal Contrasting Sex Bias in Neolithic and Bronze Age Eurasian Migrations," *Proceedings of the National Academy of Sciences USA* 114 (2017): 2657–2662.

7. I. Lazaridis and D. Reich, "Failure to Replicate a Genetic Signal for Sex Bias in the Steppe Migration into Central Europe," *Proceedings of the National Academy of Sciences USA* 114 (2017): E3873–E3874.

8. L. Saag, L. Varul, C. L. Scheib, J. Stenderup, M. E. Allentoft, L. Saag, L. Pagani, et al., "Extensive Farming in Estonia Started through a Sex-Biased Migration from the Steppe," *Current Biology* 27 (2017): 2185–2193.

9. I. Olalde, S. Mallick, N. Patterson, N. Rohland, V. Villalba-Mouco, M. Silva, K. Dulias, et al., "The Genomic History of the Iberian Peninsula over the Last 8,000 Years," *Science* 363 (2019): 1230–1234.

10. K.-G. Sjögren, I. Olalde, S. Carver, M. E. Allentoft, T. Knowles, G. Kroonen, A. W. G. Pike, et al., "Kinship and Social Organization in Copper Age Europe: A Cross-Disciplinary Analysis of Archaeology, DNA, Isotopes, and Anthropology from Two Bell Beaker Cemeteries," *PLoS ONE* 15 (2020): e0241278.

11. O. Szemerényi, "Studies in the Kinship Terminology of the Indo-European Languages, with Special Reference to Indian, Iranian, Greek and Latin," in *Acta Iranica* (Leiden: E. J. Brill, 1977), 16: 1–240.

12. B. A. Olsen, "Kin, Clan and Community in Proto-Indo-European," in *Kin, Clan and Community in Prehistoric Europe*, ed. B. Nielsen Whitehead, B. A. Olsen, and J. B. Jacquet (Copenhagen: Museum Tusculanum Press, 2019), 39–180.

13. V. M. Narasimhan, N. Patterson, P. Moorjani, N. Rohland, R. Bernardos, S. Mallick, I. Lazaridis, et al., "The Genomic Formation of South and Central Asia," *Science* 365 (2019): eeat7487.

14. D. Reich, K. Thangaraj, N. Patterson, A. L. Price, and L. Singh, "Reconstructing Indian Population History," *Nature* 461 (2009): 489–494; P. Moorjani, K. Thangaraj, N. Patterson, M. Lipson, P.-R. Loh, P. Govindaraj, B. Berger, et al., "Genetic Evidence for Recent Population Mixture in India," *American Journal of Human Genetics* 93 (2013): 422–438.

15. C. Capelli, N. Redhead, J. K. Abernethy, F. Gratrix, J. F. Wilson, T. Moen, T. Hervig, et al., "A Y Chromosome Census of the British Isles," *Current Biology* 13 (2003): 979–984.

16. M. G. Thomas, M. P. H. Stumpf, and H. Härke, "Evidence for an Apartheid-Like Social Structure in Early Anglo-Saxon England," *Proceedings of the Royal Society B* 273 (2006): 2651–2657.

17. D. Reich, N. Patterson, M. Kircher, F. Delfin, M. R. Nandineni, I. Pugach, A. M.-S. Ko, et al., "Denisova Admixture and the First Modern Human Dispersals into Southeast Asia and Oceania," *American Journal of Human Genetics* 89 (2011): 516–528.

18. M. Lipson, P.-R. Loh, N. Patterson, P. Moorjani, Y.-C. Ko, M. Stoneking, B. Berger, and D. Reich, "Reconstructing Austronesian Population History in Island Southeast Asia," *Nature Communications* 5 (2014): 4689.

19. P. Skoglund, C. Posth, K. Sirak, M. Spriggs, F. Valentin, S. Bedford, G. R. Clark, et al., "Genomic Insights into the Peopling of the Southwest Pacific," *Nature* 538 (2016): 510–513.

20. C. Posth, K. Nägele, H. Colleran, F. Valentin, S. Bedford, K. W. Kami, R. Shing, et al., "Language Continuity despite Population Replacement in Remote Oceania," *Nature Ecology and Evolution* 2 (2018): 731–740.

21. M. Lipson, P. Skoglund, M. Spriggs, F. Valentin, S. Bedford, R. Shing, H. Buckley, et al., "Population Turnover in Remote Oceania Shortly after Initial Settlement," *Current Biology* 28 (2018): 1157–1165.e7.

22. D. Reich, *Who We Are and How We Got Here: Ancient DNA and the New Science of the Human Past* (New York: Pantheon Books, 2018); P. Manning, "Migrations of Africans to the Americas: The Impact of Africans, Africa and the New World," *History Teacher* 26 (1993): 279–296.

23. P. Wade, *Race and Ethnicity in Latin America* (London: Pluto Press, 2010).

24. G. Rodriguez, *Mongrels, Bastards, Orphans, and Vagabonds: Mexican Immigration and the Future of Race in America* (New York: Pantheon Books, 2007).

25. M. Sans, T. A. Weimer, M. H. L. P. Franco, F. M. Salzano, N. Bentancor, I. Alvarez, N. O. Bianchi, and R. Chakraborty, "Unequal Contributions of Male and Female Gene Pools from Parental Populations in the African Descendants of the City of Melo, Uruguay," *American Journal of Physical Anthropology* 118 (2002): 33–44.

26. A. Gómez-Carballa, A. Ignacio-Veiga, V. Alvarez-Iglesias, A. Pastoriza-Mourelle, Y. Ruíz, L. Pineda, A. Carracedo, and A. Salas, "A Melting Pot of Multicontinental mtDNA Lineages in Admixed Venezuelans," *American Journal of Physical Anthropology* 147 (2012): 78–87.

27. M. Sandoval-Velasco, A. Jagadeesan, M. C. Ávila-Arcos, S. Gopalakrishnan, J. Ramos-Madrigal, J. V. Moreno-Mayar, G. Renaud, et al., "The Genetic Origins of Saint Helena's Liberated Africans," bioRxiv, 2020, https://doi.org/10.1101/787515.

28. M. Mörner, *Race Mixture in the History of Latin America* (New York: Little, Brown and Company, 1967).

29. Bernal Díaz del Castillo, *Historia verdadera de la conquista de la Nueva España* (Madrid: Alianza Editorial, 2016).

30. K. Bryc, C. Velez, T. Karafet, A. Moreno-Estrada, A. Reynolds, A. Auton, M. Hammer, et al., "Genome-Wide Patterns of Population Structure and Admixture among Hispanic/Latino Populations," *Proceedings of the National Academy of Sciences USA* 107 (2010): 8954–8961.

31. K. Stefflova, M. C. Dulik, A. A. Pai, A. H. Walker, C. M. Zeigler-Johnson, S. M. Gueye, R. G. Schurr, and T. R. Rebbeck, "Evaluation of Group Genetic Ancestry of Populations from Philadelphia and Dakar in the Context of Sex-Biased Admixture in the Americas," *PLoS One* 4 (2009): e7842; O. Lao, P. M. Vallone, M. D. Coble, T. M. Diegoli, M. van Oven, K. J. van der Gaag, J. Pijpe, et al., "Evaluating Self-Declared Ancestry of U.S. Americans with Autosomal, Y-Chromosomal and Mitochondrial DNA," *Human Mutation* 31 (2010): E1875–E1893.

32. A. Moreno-Estrada, S. Gravel, F. Zakharia, J. L. McCauley, J. K. Byrnes, C. R. Gignoux, P. A. Ortiz-Tello, et al., "Reconstructing the Population Genetic History of the Caribbean," *PLoS Genetics* 9 (2013): e1003925.

33. K. Bryc, E. Y. Durand, J. M. Macpherson, D. Reich, and J. L. Mountain,"The Genetic Ancestry of African Americans, Latinos and Europeans Americans across the United States," *American Journal of Human Genetics* 96 (2015): 37–53.

34. F. James Davis, *Who Is Black? One Nation's Definition* (University Park: Pennsylvania State University Press, 2001).

35. M. D. Shriver, E. J. Parra, S. Dios, C. Bonilla, H. Norton, C. Jovel, C. Pfaff, et al., "Skin Pigmentation, Biogeographical Ancestry and Admixture Mapping," *Human Genetics* 112 (2002): 387–399.

36. A. Gordon-Reed, "Sally Hemings, Thomas Jefferson and the Ways We Talk About Our Past," *New York Times*, August 24, 2017; A. Gordon-Reed, *The Hemingses of Monticello: An America Family* (New York: W. W. Norton and Company, 2009).

37. E. A. Foster, M. A. Jobling, P. G. Taylor, P. Donnelly, P. de Knijff, R. Mieremet, T. Zerjal, and C. Tyler-Smith, "Jefferson Fathered Slave's Last Child," *Nature* 396 (1998): 27–28.

38. Quoted in E. Check, "Jefferson's Descendants Continue to Deny Slave Link," *Nature* 417 (2002): 213.

Chapter 6

1. Quoted in J. T. Flynn, *God's Gold: The Story of Rockefeller and His Times* (Rahway, NJ: Quinn and Boden Company, 2007).

2. Agnar Helgason, personal communication.

3. A. Kong, "A High-Resolution Recombination Map of the Human Genome," *Nature Genetics* 31 (2002): 241–247.

4. G. Coop, "Where Did Your Genetic Ancestors Come From?," https://gcbias.org/, December 19, 2017.

5. C. Lalueza-Fox, *Genes, reyes e impostores; Una historia detectivesca tras los análisis genéticos de reyes Europeos* (Palencia: Ediciones Cálamo, 2016).

6. G. Alvarez, F. C. Ceballos, and C. Quinteiro, "The Role of Inbreeding in the Extinction of a European Royal Dynasty,"*PLos One* 4 (2009): e5174.

7. T. E. King, E. J. Parkin, G. Swinfield, F. Cruciani, R. Scozzari, A. Rosa, S.-K. Lim, et al., "Africans in Yorkshire? The Deepest-Rooting Clade of the Y Phylogeny within an English Genealogy," *European Journal of Human Genetics* 15 (2007): 288–293.

8. M. Kaufmann, *Black Tudors: The Untold Story* (Prince Frederick, MD: Highbridge Audio, 2017).

9. A. Waley, *The Secret History of the Mongols* (Cornwall, UK: House of Stratus, 2008).

10. H. Lamb, *Genghis Khan: Emperor of All Men* (Mattituck, NY: Amereon Ltd., 1986).

11. F. McLynn, *Genghis Khan: His Conquests, His Empire, His Legacy* (Boston: De Capo Press, 2016).

12. J. Pongratz, K. Caldeira, C. H. Reick, and M. Claussen, "Coupled Climate-Carbon Simulations Indicate Minor Global Effects of Wars and Epidemics on Atmospheric CO_2 between AD 800 and 1850," *Holocene* 21 (2011): 843–851.

13. J. Weatherford, *Genghis Khan and the Making of the Modern World* (New York: Broadway Books, 2005).

14. T. Zerjal, Y. Xue, G. Bertorelle, R. S. Wells, W. Bao, S. Zhu, R. Qamar, et al., "The Genetic Legacy of the Mongols," *American Journal of Human Genetics* 72 (2003): 717–721.

15. Zerjal, et al., "The Genetic Legacy of the Mongols."

16. L.-H. Wei, S. Yan, Y. Lu, S.-Q. Wen, Y.-Z. Huang, L.-X. Wang, S.-L. Li, et al., "Whole-Sequence Analysis Indicates That the Y Chromosome C2*-Star Cluster Traces Back to Ordinary Mongols, rather than Genghis Khan," *European Journal of Human Genetics* 26 (2018): 230–237.

17. T. E. King, G. G. Fortes, P. Balaresque, M. G. Thomas, D. Balding, P. M. Delser, R. Neumann, et al., "Identification of the Remains of King Richard III," *Nature Communications* 5 (2014): 5631.

18. M. H. D. Larmuseau, P. van den Berg, S. Claerhout, F. Calafell, A. Boattini, L. Gruyters, M. Vandenbosch, et al., "A Historical-Genetic Reconstruction of Human Extra-Pair Paternity," *Current Biology* 29 (2019): 4102–4107.e7.

19. L. T. Moore, B. McEvoy, E. Cape, K. Simms, and D. Bradley, "A Y-Chromosome Signature of Hegemony in Gaelic Ireland," *American Journal of Human Genetics* 78 (2006): 334–338.

20. Y. Xue, T. Zerjal, W. Bao, S. Zhu, S.-K. Lim, Q. Shu, J. Xu, et al., "Recent Spread of a Y-Chromosomal Lineage in Northern China and Mongolia," *American Journal of Human Genetics* 77 (2005): 1112–1116.

21. S. Yan, C.-C. Wang, H.-X. Zheng, W. Wang, Z.-D. Qin, L.-H. Wei, Y. Wang, et al., "Y Chromosomes of 40% of Chinese Descend from Three Neolithic Super-Grandfathers," *PLoS One* 9 (2014): e105691.

22. M. Karmin, L. Saag, M. Vicente, M. A. W. Sayres, M. Järve, U. G. Talas, S. Rootsi, et al., "A Recent Bottleneck of Y Chromosome Diversity Coincides with a Global Change in Culture," *Genome Research* 25 (2015): 459–466.

23. T. C. Zeng, A. J. Aw, and M. W. Feldman, "Cultural Hitchhiking and Competition between Patrilineal Kin Groups Explain the Post-Neolithic Y Chromosome Bottleneck," *Nature Communications* 9 (2018): 2077.

24. F. Racimo, M. Sikora, M. Vander Linden, H. Schroeder, and C. Lalueza-Fox, "Beyond Broad Strokes: Sociocultural Insights from the Study of Ancient Genomes," *Nature Review Genetics* 21, no. 6 (2020): 355–366.

Chapter 7

1. V. Galasso, "COVID: Not a Great Equalizer," *CESifo Economic Studies* 66, no. 4 (2020): 376–393; S. Galletta and T. Giommoni, "Pandemics and Inequality," VoxEU .org, October 3, 2020.

2. O. Jones, "We're about to Learn a Terrible Lesson from Coronavirus: Inequality Kills," *Guardian*, March 14, 2020.

3. S. Schifferes, "The Coronavirus Pandemic Is Already Increasing Inequality," *Conversation*, April 10, 2020; O. Khan, "Coronavirus Exposes How Riddled Britain Is with Racial Inequality," *Guardian*, April 20, 2020.

4. M. Sacchetti, "'I'm Scared': Black People—Many of the Immigrants—Make Up Less Than 2 Percent of Maine's Population but Almost a Quarter of Its Coronavirus Cases," *Washington Post*, July 30, 2020.

5. J. Kirby, "What We Can Learn from the 'Second Wave' of Coronavirus Cases in Asia," *Vox*, April 17, 2020.

6. L. Wade, "From Black Death to Fatal Flu, Past Pandemics Show Why People on the Margins Suffer Most," *Science*, May 14, 2020.

7. D. Quintero, "A History of Epidemics: In Past, Navajos Survived Many Epidemics. Spanish Flu, Virus Pose Danger," *Navajo Times*, April 19, 2020.

8. P. Ralph and G. Coop, "The Geography of Recent Genetic Ancestry across Europe," *PLoS Biology* 11 (2013): e1001555.

9. M. Ferrando-Bernal, C. Morcillo-Suarez, T. de-Dios, P. Gelabert, S. Civit, A. Díaz-Carvajal, I. Ollich-Castanyer, et al., "Mapping Co-Ancestry Connections between the Genome of a Medieval Individual and Modern Europeans," *Scientific Reports* 10 (2020): 6843.

10. G. Palsson, *The Man Who Stole Himself: The Slave Odyssey of Hans Jonathan* (Chicago: University of Chicago Press, 2016).

11. N. Wolchover, "What Will Future Humans Look Like?," *LiveScience*, September 18, 2012.

12. M. L. Antonio, Z. Gao, H. M. Moots, M. Lucci, F. Candilio, S. Sawyer, V. Oberreiter, et al., "Ancient Rome: A Genetic Crossroads of Europe and the Mediterranean," *Science* 366 (2019): 708–714.

13. D. Reich, personal communication.

14. A. Huxley, *Brave New World* (New York: Random House, 2004).

15. D. Cyranoski, "The CRISPR-Baby Scandal: What's Next for Human Gene-Editing," *Nature* 566 (2019): 440–442.

16. J. Cohen, "The Untold Story of the 'Circle of Trust' behind the World's First Gene-Edited Babies," *Science*, August 1, 2019.

17. S. Sigal, "A Celebrity Biohacker Who Sells DIY Gene-Editing Kits Is under Investigation," *Vox*, May 19, 2019.

18. Quoted in S. Marsh, "Extreme Biohacking: The Tech Guru Who Spent $250,000 Trying to Live for Ever," *Guardian*, September 21, 2018.

19. Y.-P. Tang, E. Shimizu, G. R. Dube, C. Rampon, G. A. Kerchner, M. Zhuo, G. Liu, and J. Z. Thsien, "Genetic Enhancement of Learning and Memory in Mice," *Nature* 401 (1999): 63–69.

20. H. Elberg, J. Troelsen, M. Nielsen, A. Mikkelsen, J. Mengel-From, K. W. Kjaer, and L. Hansen, "Blue Eye Color in Humans May Be Caused by a Perfectly Associated Founder Mutation in a Regulatory Element Located within the HERC2 Gene Inhibiting OCA2 Expression," *Human Genetics* 123 (2008): 177; J. L. Rees, "The Melanocortin 1 Receptor (MC1R): More Than Just Red Hair," *Pigment Cell Research* 13 (2000): 135–140.

21. D. J. Galton, *Eugenics: The Future of Human Life in the 21st Century* (London: Abacus, 2002).

22. F. E. Vizcarrondo, "Human Enhancement: The New Eugenics," *Linacre Quarterly* 81 (2014): 239–243.

23. R. Vilas, F. C. Ceballos, L. Al-Soufi, R. González-García, C. Moreno, M. Moreno, L. Villanueva, et al., "Is the 'Habsburg Jaw' Related to Inbreeding?," *Annals of Human Biology* 46 (2019): 553–561.

24. F. Morton, *The Rothschilds: A Family Portrait* (New York: Diversion Books, 2014).

25. G. M. Petersen, J. I. Rotter, R. M. Cantor, L. L. Field, S. Greenwald, J. S. Lim, C. Roy, et al., "The Tay-Sachs Disease Gene in North American Jewish Populations: Geographic Variations and Origin," *American Journal of Human Genetics* 35 (1983): 1258–1269.

26. K. A. Strauss, E. G. Puffenberger, and H. Morton, "One Community's Effort to Control Genetic Disease," *American Journal of Public Health* 102 (2012): 1300–1306.

27. C. Deppen, "Genetic Disease Is Ravaging Lancaster County's Amish, and Helping to Change Medicine for All of Us," *Pennsylvania Real-Time New*, January 5, 2019.

28. J. W. von Goethe, *Maxims and Reflections of Goethe,* trans. Bailey Saunders (New York: Macmillan Company, 1906; Project Gutenberg, 2010), e-book #33670, sec. III, line 138, https://www.gutenberg.org/files/33670/33670-h/33670-h.htm.

29. T. Rai and W. Wible, "In Flux and under Threat," *Science* 369 (2020): 1174–1175.

Index